In All Probability

Investigations in Probability and Statistics

TEACHER'S GUIDE

Grades 3–6

Skills
Recording Data, Organizing Data, Analyzing Data, Graphing, Working with a Partner, Predicting, Describing, Comparing, Estimating, Drawing Conclusions

Concepts
Probability, Statistics, Prediction

Themes
Patterns of Change, Models and Simulations

WHAT ARE THEMES?
Themes can be seen as major, recurring ideas that provide a framework for the science curriculum. For more on what GEMS means by themes, please see page . **vi**

Mathematics Strands
Probability and Statistics, Logic, Pattern, Number

Time
Twelve or more 45-60 minute sessions

Celia Cuomo

LHS GEMS

Great Explorations in Math and Science
Lawrence Hall of Science
University of California at Berkeley

Illustrations
Carol Bevilacqua
Lisa Klofkorn

Photographs
Richard Hoyt

Lawrence Hall of Science, University of California,
Berkeley, CA 94720

Chairman: Glenn T. Seaborg
Director: Marian C. Diamond

> **Publication of *In All Probability* was made possible by a grant from the McDonnell Douglas Foundation and the McDonnell Douglas Employees Community Fund. The GEMS Project and the Lawrence Hall of Science greatly appreciate this support.**

Initial support for the origination and publication of the GEMS series was provided by the A.W. Mellon Foundation and the Carnegie Corporation of New York. GEMS has also received support from the McDonnell-Douglas Foundation and the McDonnell-Douglas Employees Community Fund, the Hewlett Packard Company Foundation, and the people at Chevron USA. GEMS also gratefully acknowledges the contribution of word processing equipment from Apple Computer, Inc. This support does not imply responsibility for statements or views expressed in publications of the GEMS program. Under a grant from the National Science Foundation, GEMS Leader's Workshops have been held across the country. For further information on GEMS leadership opportunities, or to receive a publication brochure and the *GEMS Network News*, please contact GEMS at the address and phone number below.

International Standard Book Number: 0-912511-83-4

COMMENTS WELCOME

Great Explorations in Math and Science (GEMS) is an ongoing curriculum development project. GEMS guides are revised periodically, to incorporate teacher comments and new approaches. We welcome your criticisms, suggestions, helpful hints, and any anecdotes about your experience presenting GEMS activities. Your suggestions will be reviewed each time a GEMS guide is revised. Please send your comments to: GEMS Revisions, c/o Lawrence Hall of Science, University of California, Berkeley, CA 94720. The phone number is (510) 642-7771.

Great Explorations in Math and Science (GEMS) Program

The Lawrence Hall of Science (LHS) is a public science center on the University of California at Berkeley campus. LHS offers a full program of activities for the public, including workshops and classes, exhibits, films, lectures, and special events. LHS is also a center for teacher education and curriculum research and development.

Over the years, LHS staff have developed a multitude of activities, assembly programs, classes, and interactive exhibits. These programs have proven to be successful at the Hall and should be useful to schools, other science centers, museums, and community groups. A number of these guided-discovery activities have been published under the Great Explorations in Math and Science (GEMS) title, after an extensive refinement process that includes classroom testing of trial versions, modifications to ensure the use of easy-to-obtain materials, and carefully written and edited step-by-step instructions and background information to allow presentation by teachers without special background in mathematics or science.

Staff

Glenn T. Seaborg, **Principal Investigator**
Jacqueline Barber, **Director**
Kimi Hosoume, **Assistant Director**
Cary Sneider, **Curriculum Specialist**
Carolyn Willard, **GEMS Centers Coordinator**
Laura Tucker, **GEMS Workshop Coordinator**
Katharine Barrett, Kevin Beals, Ellen Blinderman, Beatrice Boffen, John Erickson, Jaine Kopp, Laura Lowell, Linda Lipner, Debra Sutter, Rebecca Tilley,
        Staff Development Specialists
Jan M. Goodman, **Mathematics Consultant**
Cynthia Eaton, **Administrative Coordinator**
Karen Milligan, **Distribution Coordinator**
Lisa Haderlie Baker, **Art Director**
Carol Bevilacqua, Rose Craig and Lisa Klofkorn, **Designers**
Lincoln Bergman, **Principal Editor**
Carl Babcock, **Senior Editor**
Kay Fairwell, **Principal Publications Coordinator**
Erica De Cuir, Nancy Kedzierski, Felicia Roston, Vivian Tong, Stephanie Van Meter, Mary Yang, **Staff Assistants**

Contributing Authors

Jacqueline Barber	Jean Echols
Katharine Barrett	Jan M. Goodman
Kevin Beals	Alan Gould
Lincoln Bergman	Kimi Hosoume
Celia Cuomo	Susan Jagoda
Philip Gonsalves	Larry Malone
Jaine Kopp	Cary I. Sneider
Linda Lipner	Debra Sutter
Laura Lowell	Jennifer Meux White
Linda De Lucchi	Carolyn Willard

Reviewers

We would like to thank the following educators who reviewed, tested, or coordinated the reviewing of *this series* of GEMS materials in manuscript and draft form. Their critical comments and recommendations, based on presentation of these activities nationwide, contributed significantly to these GEMS publications. Their participation in the review process does not necessarily imply endorsement of the GEMS program or responsibility for statements or views expressed. Their role in an invaluable one, and their feedback is carefully recorded and integrated as appropriate into the publications. **THANK YOU!**

ALASKA
Coordinator: **Cynthia Dolmas Curran**

Iditarod Elementary School, Wasilla
Cynthia Dolmas Curran
Jana DePriest
Christina M. Jencks
Abby Kellner-Rode
Beverly McPeek

Sherrod Elementary School, Palmer
Michael Curran
R. Geoffrey Shank
Tom Hermon

CALIFORNIA
GEMS Center, Huntington Beach
Coordinator: **Susan Spoeneman**

College View School, Huntington Beach
Kathy O'Steen
Robin L. Rouse
Karen Sandors
Lisa McCarthy

John Eader School, Huntington Beach
Jim Atteberry
Ardis Bucy
Virginia Ellenson

Issac Sowers Middle School, Huntington Beach
James E. Martin

San Francisco Bay Area
Coordinator: **Cynthia Eaton**

Bancroft Middle School, San Leandro
Catherine Heck
Barbara Kingsley
Michael Mandel
Stephen Rutherford

Edward M. Downer Elementary School, San Pablo
M. Antonieta Franco
Nancy Hirota
Barbara Kelly
Linda Searls
Emily Teale Vogler

Malcolm X Intermediate School, Berkeley
Candyce Cannon
Carole Chin
Denise B. Lebel
Rudolph Graham
DeEtte LaRue
Mahalia Ryba

Marie A. Murphy School, Richmond
Sally Freese
Dallas Karahalios
Susan Jane Kirsch
Sandra A. Petzoldt
Versa White

Marin Elementary School, Albany
Juline Aguilar
Chris Bowen
Lois B. Breault
Nancy Davidson
Sarah Del Grande
Marlene Keret
Juanita Rynerson
Maggie J. Shepherd
Sonia Zulpo

Markham Elementary School, Oakland
Alvin Bettis
Eleanor Feuille
Sharon Kerr
Steven L. Norton
Patricia Harris Nunley
Kirsten Pihlaja
Ruth Quezada

Martin Luther King, Jr. Jr. High
Phoebe Tanner

Sierra School, El Cerrito
Laurie Chandler
Gary DeMoss
Tanya Grove
Roselyn Max
Norman Nemzer
Martha Salzman
Diane Simoneau
Marcia Williams

Sleepy Hollow Elementary School, Orinda
Lou Caputo
Marlene Fraser
Carolyn High
Janet Howard
Nancy Medbery
Kathy Mico-Smith
Anne H. Morton
Mary Welte

Walnut Heights Elementary School, Walnut Creek
Christl Blumenthal
Nora Davidson
Linda Ghysels
Julie A. Ginocchio
Sally J. Holcombe
Thomas F. MacLean
Elizabeth O'Brien
Gail F. Puleo

Willard Junior High School, Berkeley
Vana James
Linda Taylor-White
Katherine Evans

GEORGIA
Coordinator: **Yonnie Carol Pope**

Dodgen Middle School, Marietta
Linda W. Curtis
Joan B. Jackson
Marilyn Pope
Wanda Richardson

Mountain View Elementary School,
Marietta
Cathy Howell
Diane Pine Miller
Janie E. Stokes
Elaine S. Toney

NEW YORK
Coordinator: **Stanley J. Wegrzynowski**

Dr. Charles R. Drew Science Magnet,
Buffalo
Mary Jean Syrek
Renée C. Johnson
Ruth Kresser
Jane Wenner Metzger
Sharon Pikul

Lorraine Academy, Buffalo
Francine R. LoGrippo
Clintonia R. Graves
Albert Gurgol
Nancy B. Kryszczuk
Laura P. Parks

OREGON
Coordinator: **Anne Kennedy**

Myers Elementary School, Salem
Cheryl A. Ward
Carol Nivens
Kent C. Norris
Tami Socolofsky

Terrebonne Elementary School,
Terrebonne
Francy Stillwell
Elizabeth M. Naidis
Carol Selle
Julie Wellette

Wallowa Elementary School, Wallowa
Sherry Carman
Jennifer K. Isley
Neil A. Miller
Warren J. Wilson

PENNSYVANIA
Coordinator: **Greg Calvetti**

Aliquippa Elementary School,
Aliquippa
Karen Levitt
Lorraine McKinin
Ted Zeljak

Duquesne Elementary School,
Duquesne
Tim Kamauf
Mike Vranesivic
Elizabeth A. West

Gateway Upper Elementary School,
Monroeville
Paul A. Bigos
Reed Douglas Hankinson
Barbara B. Messina
Barbara Platz
William Wilshire

Ramsey Elementary School,
Monroeville
Faye Ward

WASHINGTON
Coordinator: **David Kennedy**

Blue Ridge Elementary School, Walla
Walla
Peggy Harris Willcuts

Prospect Point School, Walla Walla
Alice R. MacDonald
Nancy Ann McCorkle

MORE ON THEMES

The word "themes" is used in many different ways in both ordinary usage and in educational circles. In the GEMS series, themes are seen as key recurring ideas that cut across all the scientific disciplines. Themes are bigger than facts, concepts, or theories. They link various theories from many disciplines. They have also been described as "the sap that runs through the curriculum," to convey the sense that they permeate through and arise from the curriculum. By listing the themes that run through a particular GEMS unit on the title page, we hope to assist you in seeing where the unit fits into the "big picture" of science, and how the unit connects to other GEMS units. The theme "Patterns of Change," for example, suggests that the unit or some important part of it exemplifies larger scientific ideas about why, how, and in what ways change takes place, whether it be a chemical reaction or a caterpillar becoming a butterfly. GEMS has selected 10 major themes:

Systems & Interactions	**Scale**
Models & Simulations	**Structure**
Stability	**Energy**
Patterns of Change	**Matter**
Evolution	**Diversity & Unity**

If you are interested in thinking more about themes and the thematic approach to teaching and constructing curriculum, you may wish to obtain a copy of our handbook, *To Build A House: GEMS and the Thematic Approach to Teaching Science.* For more information and an order brochure, write or call GEMS, Lawrence Hall of Science, University of California, Berkeley, CA 94720. (510) 642-7771. **Thanks for your interest in GEMS!**

Table of Contents

Acknowledgments

Special thanks goes to:

• Nancy Hirota and her 5th and 6th grade students at Downer Elementary School, for participating enthusiastically in the original documentation of the activities and providing very helpful and encouraging feedback.

• All the teachers and students who field-tested this guide, gathered data, wrote comments, drew graphs and shared their ideas for modifications and improvements. (Teachers and schools who took part in local and national testing are listed in the front of this guide.)

• Aggie Brenneman, Catherine Keyes and their students from Park Day School in Oakland, who generously allowed us to photograph classroom activities.

• Kimi Hosoume, GEMS Assistant Director and my knowledgeable "GEMS buddy," who provided invaluable support as she guided me through the work of transforming these activities into a finished book.

• Jan Goodman and Jaine Kopp, two of my colleagues in the Mathematics Education Program, for developing and enhancing many of the games in this unit, and Linda Lipner, the Director of the Mathematics Education Program, for her contributions to the revision process.

• Cynthia Ashley, Cynthia Eaton, and Nancy Kedzierski of the GEMS staff, who facilitated the many and varied logistics required to test and produce this guide.

• Scarlett Manning for the lively and colorful cover, and Carol Bevilacqua for the artwork and design inside.

• Carl Babcock, Richard Hoyt and Laurence E. Bradley for the photographs that capture young people actively and enjoyably engaged in mathematical learning.

• Lincoln Bergman for the poem that appears on the back cover.

Introduction

Your students experience probability and statistics frequently in their daily lives. They play games with spinners, coins, cards, or dice, and all of these involve probability. They encounter statistics on food packaging, on television commercials, in the sports page, on baseball cards, in political opinion polls, or on a graph of the school fund-raising drive.

The activities in this unit engage your students in active mathematics situations that are fun, exciting and allow students to further develop mathematics literacy, especially relating to statistics, graphing and probability. Helping students understand how often and in how many ways their lives and decisions relate to probability and statistics can be one of the most lasting goals of this unit. Such knowledge will be very important to them as citizens and decision-makers, whether it be making an informed consumer choice, evaluating economic statistics in newspaper articles, judging the chances of winning a lottery, or any of hundreds of other moments in daily life.

It is highly recommended that you read the entire "Behind the Scenes" background section. For more information on the many ways we encounter statistics and probability in our lives, see "Who Uses Statistics and Probability?" on page 72.

If you and your students use a textbook for most math lessons, you will find this guide a welcome change from "traditional math." There are no pages of equations or word problems. Please rest assured, however, that this does not mean that there is little solid mathematics in these activities—on the contrary! Throughout these activities, students actually use and apply mathematics knowledge and skills in many highly motivating ways. Within the context of conducting probability investigations, students generate data for themselves, graph, interpret data, and communicate (by speaking to one another and to you, by conveying information in a graph, and by writing). Students also predict, describe, compare, compute and draw conclusions. They apply numerous practical and logical thinking skills and gain insight into how the major mathematical ideas of probability and statistics relate to them and their world. These key mathematics concepts and thinking skills will also serve the children in other areas of learning, such as in science, social studies and the language arts.

*In communities where dice may be offensive due to their association with gambling, the euphemism "probability cubes" may be substituted for dice. Labeling cubes with numbers rather than dots may also lessen the association with gambling, as does the use of non-traditional dice that may be shapes other than cubes. On a larger scale, since penny flipping, horse races, game sticks, and many other games and sports can also involve gambling, some teachers may want to discuss the social /ethical aspects of gambling more directly from a social studies vantagepoint. Other possible and interesting extensions might be: to explore the probability involved in a game like Bingo, which has become an important source of revenue for some Indian nations; or to explore the probability concepts and chances of winning involved in statewide lottery games, which help fund education in some states. How about the direct mail magazine promotions that assure us millions of dollars, **if** we return the winning number? Providing students with basic tools to think through these "odds" is very **likely** to stand them in good stead throughout their lives!*

In Activity 1: Penny Flip, students toss pennies as they investigate the probability of getting heads and tails, and develop ideas to describe the results they observe. In **Activity 2: Track Meet**, students play a game with two different spinners (one fair and one unfair) and compare the results. In **Activity 3: Roll a Die**, students toss one die to generate data in preparation for **Activity 4: Horse Race**, in which they play a game with two dice and investigate surprising results. In a creative finale, **Activity 5: Native American Game Sticks**, introduces students to a version of a Native American game of chance. After making their own colorfully-designed sticks, students use them to play the game.

The games and activities included in this guide are rich with opportunity for students in Grades 3 through 6, and have been used successfully with both younger and older students. You can make the best judgments about how to present the activities based on the ability and experience of your students.

Generally, students at Grades 3 and 4 focus more on **recording** and **representing** data, and investigate answers to important questions such as: How can we organize our results? What information can we understand from the graph? At this introductory stage, even though students may not yet have an understanding of the probability concepts involved, it is still important to ask them **why** they think the class got certain results. Early experiences such as these build a concrete and intuitive foundation upon which to build deeper understandings of probability that will develop in later grades.

In general, students in Grades 5 and 6 can concentrate more on probability aspects once the data has been recorded and graphed. Supplied with hands-on activities, opportunities for discussion with each other, and the time to make comparisons and write down their reasoning, many students in these upper elementary years are able to begin to understand probability. For example, your students may see that, given enough flips of a penny, heads will land up *about* half the time. Even though the chance of getting heads when you flip a penny is 1 out of 2 (or 50%), this is only approximately what you will get, not exactly what you will get every time.

For further information about the concept of probability, please see "Probability" beginning on page 67 and "The Probability in the Activities" on page 74.

Each activity involves your students in collecting, organizing and interpreting data, usually by recording on a data sheet and then making a graph. Data sheets are included for those activities that require them, or students can make their own. Additional, removable copies of data sheets are provided at the back of the book for copying convenience. Two kinds of graph paper are also provided. It's important to emphasize that most students, including third graders, are capable of constructing their own data sheets and making their own graphs, and can learn very valuable lessons by doing so. Many teachers have found positive changes in their students abilities to construct and interpret data after they have had the challenge of making their own data sheets and graphs. The data sheets and graphs are provided in this guide to allow maximum flexibility for teachers in choosing an approach that is best for their individual classrooms.

One teacher wrote: "My students' graphs have become more organized and clearer over the course of doing these activities, with more labeling spontaneously occurring in each successive activity."

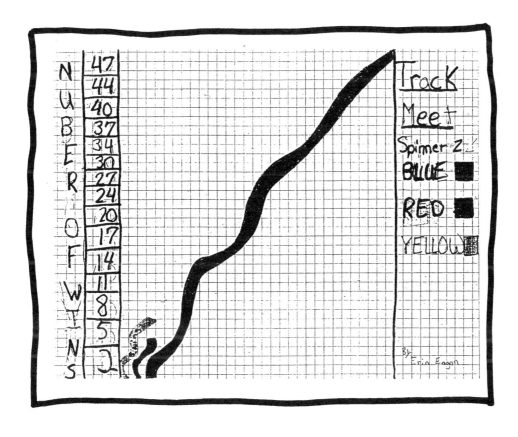

Graphing ideas are included with each activity, and more detailed information is provided in "Representing Data" on page 69.

Quiet writing time allows students to reflect on the events or discoveries of the lesson, recast their learning into their own words, and clarify their thoughts as they transfer their ideas to paper. Some teachers have found a place in the school day for writing in math by using a math-related question or "starter" during writing time. Several writing suggestions are included with each activity, and you are likely to think of others as you and your students proceed through this guide.

For more information about using writing in conjunction with mathematics, please see "Writing and Mathematics" on page 73.

Summary Outlines for all activities are provided, starting on page 83, for quick reference and to assist you in guiding your students through these activities in an organized fashion.

As you present the activities in this guide, look for evidence (which we're sure you'll find!) that students are curious, make predictions, ask questions, interpret graphs, and suggest other probability experiments to try. The chances are great that, when you provide opportunities for your students to experiment with probability in several forms, encourages them to consider and analyze what is happening, and ask them thought-provoking, open-ended questions–you are setting up an environment where learning can, *in all probability*, flourish.

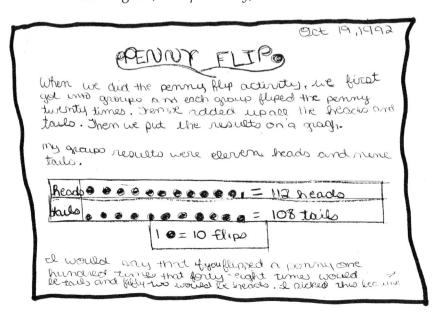

Time Frame

The following time frame, derived from results of local and national trial testing, can serve as a general guideline for your planning. Please note, however, that these are approximations, and in no way hard-and-fast schedules. You are the best judge of what works best with your students, and how much time to spend on various portions of an activity and its related discussions. Individual teachers vary greatly in how much time they spend, depending on class dynamics and their own plans/preferences. Some of the trial test teachers spent considerably longer than the times given below. Often teachers decide to extend the learning over more sessions than indicated, perhaps playing the game on three or four different days for fifteen or twenty minutes, then spending several sessions on graphing and making conclusions.

Activity 1: Penny Flip
Session 1: Heads or Tails? ....45–60 minutes
Session 2: Graphing and Class Conclusions ....45–60 minutes

Activity 2: Track Meet
Session 1: Track Meet—(Spinner 1) ....45–60 minutes
Session 2: Track Meet Rematch—(Spinner 2 or 3) ....45–60 minutes
Session 3: Graphing and Class Conclusions ....45-60 minutes

Activity 3: Roll a Die
Session 1: Gathering Data ....45–60 minutes
Session 2: Graphing and Class Conclusions...........................45–60 minutes

Activity 4: Horse Race
Session 1: Racing the Horses ....45–60 minutes
Session 2: How Many Combinations? ....45–60 minutes

Activity 5: Native American Game Sticks
Session 1: Making and Playing Game Sticks ....two class periods
                                                                                        of 45–60 minutes
Session 2: A Probability Experiment ....45–60 minutes

Activity 1: Penny Flip

Overview

In this activity, students explore the probability of getting heads and tails when they flip a penny. The sequence of this activity establishes a method of investigation that will be followed (with some modifications) in many of the other probability activities in this unit:

First, the students **make predictions** and **watch a demonstration** of how to generate and record the data.

Next, students **generate and record their own data in pairs, and discuss the results with the class**; then, students **devise a method of collecting and displaying data** from the whole class; and finally students **analyze the results orally, graphically and in writing.**

For younger students, the tasks of generating, collecting and recording data are the major tasks. For older students, once they have the data generated, recorded and organized, you can help them begin to look at the concept of *probability* by asking questions to aid them in thinking analytically about their results.

What You Need

For the class:
❑ 1 transparency for recording data ("Penny Flip" data sheet, graph paper, or blank transparency)
❑ 1 penny
❑ 1 overhead projector
❑ 1 overhead pen
❑ sponge or paper towels to clean projector
❑ writing paper
❑ graph paper, one per student, for individual graphs
    OR
❑ butcher paper or large graph paper, for a class graph

For each pair of students:
❑ 1 penny
❑ 1 "Penny Flip" Data Sheet (page 16), graph paper or blank paper
❑ crayons, markers or colored pencils

A fifth grade teacher was unsure at first that this activity would be challenging enough for her students. She commented, "I thought my students would consider this way below their level and that the 50-50 results would be obvious to them. Wrong! They flipped their pennies and recorded the data with gusto!"

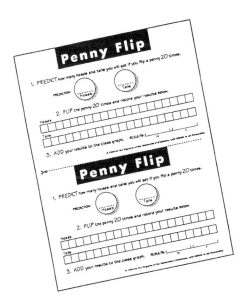

Wide rolls of graph paper are available from some teaching supply stores and catalogs. Easel pads with graph paper are available at some office supply stores.

Getting Ready

1. If your students are not used to making predictions, talk with them about the nature of predicting and give them some practice. See "Wild Guesses and Educated Predictions" on page 68.

2. Decide how you will have your students record the data they generate with their partner in Session 1. If they will use the data sheet provided or graph paper, make enough copies for each pair to have one and make an additional copy on a transparency. If students will be using blank paper for their data sheets, you will need a blank transparency.

3. For more suggestions on graphing with students, see "Representing Data" on page 69.

4. Have students begin collecting graphs from newspapers and magazines. These will be used in Session 3 of Activity 2: Track Meet. For the class, you will need at least as many graphs as students in your class.

Some teachers have students use scrap paper for recording individual results. Other teachers have students record the data each pair obtains in a math journal.

Session 1: Heads or Tails?

Students watch as you demonstrate the penny flip and how to record the results. They flip their pennies, record their results and discuss and compare their findings with the rest of the class. Some students may have already flipped coins and collected data. That's fine, because they will have that much more experience to bring to the activity! Great value can be gained from doing something more than once, in a different way, and at a later age.

Introducing the Penny Flip

1. Encourage students to share what they know about coin flipping. What are the possible outcomes? Why do people flip coins? What are some ways to flip a coin?

2. Tell students that they will flip pennies in this activity. Ask students to predict how many heads and tails they think someone would get if a penny were flipped twenty times.

3. Record these predictions on the board. If your students need the practice, you can ask them for all the other sums of twenty, and ask if any of them seem very unlikely to occur. For example, it would be very unlikely to get 19 "heads" and one "tails."

4. Have students explain how they came up with their prediction or guess. They may have made a wild guess, think that "tails never fails," or have other reasons.

5. Show the "Penny Flip" data sheet on the overhead projector or ask students how to record the results of the penny tosses. On the overhead, demonstrate one or two of the methods they suggest, using either a blank transparency or the transparency of the graph paper. Remind students there are other ways to record data in addition to what you are showing them.

6. Demonstrate flipping a penny and recording the data. Choose a student to flip the penny. Model how to record the results on the transparency data sheet.

Note: Even if your students seem familiar with the technique and procedure, they are likely to be quite interested in the results. An active presentation of instructions, with student participation, will help keep students involved and better prepared for the individual or pair activity to follow.

7. Stop two or three times during the 20 flips so the class can look at the data accumulated. Ask questions, such as, "How many times has the student flipped the penny?" "How many more flips to reach 20?" "What do the results show so far?" **An important part of the prediction process is modifying a prediction as more information is obtained. Students can change their predictions as the penny flip progresses. That is not "cheating," but making use of additional information.**

8. After twenty tosses are completed, ask students for "true statements" about the graph. For example, if there are 8 heads and 12 tails, responses might include: "There are fewer heads than tails." "There are four more tails than heads." Take a look at the predictions on the board. Ask, "Were the results close to your guess?" "Would you change your prediction next time? Why?"

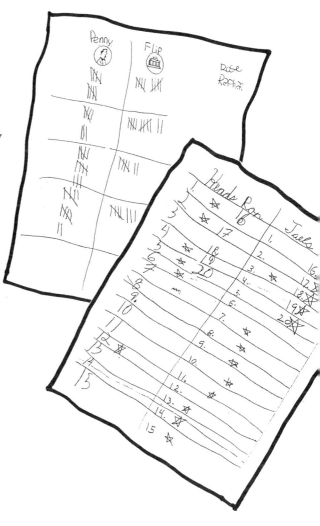

*It is easy for students to become so caught up in the task that they forget to keep track of how many times they have flipped the penny. You may need to remind them to use tally marks to record each flip. Students may make some statements that are inferences, rather than direct observations from the graph. You may want to emphasize an important distinction, i.e., **"true statements"** are facts, direct observations from the graph, for example: This time, we got more tails than heads. Examples of **inferences** are: Tails cheated. Tails will always win, etc. Some students may be familiar with the distinction between facts and inferences from social studies or language arts.*

The Penny Flip

1. Tell the students that they will have an opportunity to flip a penny with a partner to gather data. Ask students for ways they could share responsibilities of flipping and recording. Older students may bring up the idea of being consistent in the way the penny will be flipped. If so, the class can decide how to flip the penny.

2. Distribute either the "Penny Flip" data sheet or materials for students to make their own recording sheets. Have students predict how many heads and tails will come up in 20 flips and record their predictions on the data sheet.

3. Distribute one penny to each pair. Remind them about keeping track of their flips and to stop after 20. Have students generate data by flipping the penny, and recording their results. Encourage them to record data accurately.

Some teachers have suggested using a mat of paper, felt, or cardboard to muffle the sound of the pennies flipping.

4. After students are finished and have put aside the pennies, invite them to report their data to the class. Students should state their results and say whether or not their prediction was close. Ask students if they were surprised by their results and if so why. The concepts of luck, chance, and fairness may be brought up by the students—encourage them to discuss their ideas.

5. Have students give "true statements" about their data, and ask others to raise their hands when they hear a statement that is true for their own data. "True statements" can include:

> "We got 13 heads."
> "We got 12 tails."
> "We got the same number of heads and tails."
> "We got more heads than tails."

Some students may want to talk about their lucky penny or the special way they flip that makes the coin come up heads or tails. Let students tell about and demonstrate their technique, and then challenge them to conduct a separate flipping test at a later time to test their ideas.

6. You may want students to flip their pennies another 20 (or more) times in order to have more data to analyze. Ask students if they think they will get the same results during the next round of data gathering and why they think so.

7. Save student data sheets to use in the next session.

Session 2: Graphing and Class Conclusions

In this session, students graph the results and discuss the graphs. They also write about the penny flip activity. Both writing and graphing take time and concentration on the part of students, so allow ample time.

 Graphing and Discussing

1. Return the "Penny Flip" data sheets to the students. Ask for students to review the activity and the results.

2. Survey students to see how many think there will be more heads, more tails, or an equal number of heads and tails for the entire class. Ask students to predict how many heads and tails they think the class would have all together, if they counted everyone's results.

3. Gather data from all groups and record it so all can see it. Provide students with ample time to do the addition and check their answers. Resolve any discrepancies, so that all agree with the final totals. Younger students can combine their results with those of one or two other groups to keep the numbers smaller, or each group can post their results on a large class graph, in order to model a way of representing data.

In commenting about the time needed to do the graphing and writing sessions, one teacher said, "Frustration with time constraints changed to pure excitement in watching my kids learn. As the unit progressed, I said—to heck with the time."

4. Ask questions to elicit suggestions from students about such things as: information to be included be on the graph, how to fit all the data onto the paper, labels needed on each axis, and a descriptive title.

5. Depending on your students' previous graphing experience, you may or may not want to demonstrate a method of graphing on the overhead before they begin their work.

6. Students are likely to need a class period to complete their graphs.

7. Afterward, ask for "true statements" about their graphs. Responses will probably include sentences such as: "Our class flipped pennies 320 times altogether." "The two lines on the graph are almost the same length." "We got nearly the same number of heads as tails." "We got 165 heads."

8. You may want to record these statements on the overhead or the chalkboard as examples for students, to use later when writing about the investigation.

9. Ask students to tell why they think the class got the results they did. You may hear some surprising explanations from your students, but you'll also gain a clearer picture of their conceptions about probability, chance and luck.

10. Have students think silently for a moment, and make a prediction about how many heads and tails they would get if they flipped a penny 100 times. Have a few students report their predictions to the class. Ask how they could use their class data to help make a prediction about the next 100 tosses of a penny? Point out that the students now have information or *data* to rely on, and they can take that into account as they make predictions, to make their predictions more accurate. In this way, people can make **"better" or "best" or "educated guesses"** and improve their predictions by taking into account prior information and experience.

For Older Students

1. Ask students if they know what *probability* is. Begin explaining the term probability as the **chance** that a certain event will happen, such as getting "tails" when you flip a penny. (Students are generally familiar with the word "chance," which is sometimes used to mean probability, as in "there is a 30% chance of rain tomorrow.") Ask when they have heard the words *chance* or *probability* used.

2. Explain that one of the things that mathematicians and other scientists do is investigate the chance or likelihood that something will or will not happen. This is called the study of probability, or the *theory of probability*. According to probability theory, as relates to the penny flip, there is an **equally likely chance** of getting either heads or tails each time you flip a coin.

3. Probability tells us that, if a penny is flipped many times, it is *likely* that it will come up about half heads and half tails. You may want to ask older students to compare the results of their work with partners (20 flips) with the results of the class as a whole. Emphasize that **probability tells us approximately, but not exactly, what will happen.**

Do not be concerned if students "don't get it," for example, if they still persist in believing that heads/tails do not each occur about half the time. What is important is that your students are thinking, conjecturing, and speculating. Further understanding will come as they gain more experience, have more opportunities to discuss and compare, and as they mature developmentally.

For younger students, it is developmentally appropriate for them to concentrate on generating data, collecting and organizing it, making graphs and interpreting graphs. You may also want to introduce some of the ideas below, to stimulate thought and discussion, depending on the experience of your students.

The phrase "equally likely chance" will occur in other activities in this guide. It is a helpful and fundamental concept for children to begin to consider.

This is a good time to see how the numerals in a fraction represent real things. When flipping one penny, the chance of getting tails is 1/2. The denominator, the "2," represents the total possibilities—heads and tails. The numerator, the "1," represents the possibility that we are looking at, in this case heads. If students are working with decimals, they can translate the fractional probability to a decimal. There will be opportunities in other activities to return to fractional and decimal representations of probability.

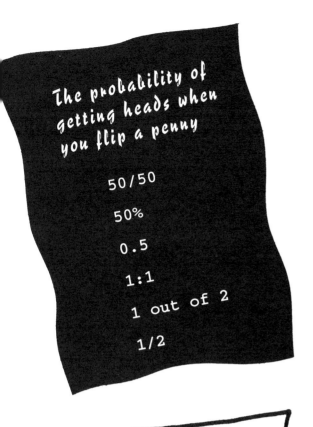

The probability of getting heads when you flip a penny

50/50

50%

0.5

1:1

1 out of 2

1/2

Penny flip

When we did the penny flip we got with a partner and flipped the penny 20 times. Then you recorded your results. Then we got together as a class and added up all the results and made a graph. My partner was Stephanie. Our results were 9 heads and 11 tails. We got the exact opposite from what we estimated. The class together got 112 heads and 108 tails. The graph we made wasn't big enough so 1 square. ad to equal 4 flips.

A teacher commented: "Keeping math journals was a wonderful idea, because the kids could look back on what they did. I also got a sense of what students were really picking up, especially those who tend to be less outspoken in group discussions."

4. You can introduce symbolic ways for writing probabilities as appropriate to your students. For example, when flipping a penny, there is a 50/50 probability of getting heads. This can also be represented in other ways: a 50% probability of getting heads; a 1 out of 2 probability of getting heads; the chance of getting heads is 1/2; the chance of getting heads is 0.5. Still another way of writing this is as a 1:1 ratio of heads to tails.

Writing

Writing suggestions are included with each activity in this guide. Before having students write, they can share ideas in collaborative groups to help stimulate their thinking. You many also want to ask them about the role of writing, for example: when have they seen people writing, why people write, why people keep notes or journals.

• Have students describe the "Penny Flip" investigation in a journal. Have them record their own results, as well as the class results and graph. Encourage them to include drawings or diagrams along with their written description.

• Have students make a prediction about what would happen if they were to flip a penny 100 times, and write about why they made that prediction.

Going Further

1. Do more trials of penny flipping, perhaps until the class has flipped 1000 times. Analyze the results. Compare your class data with another class who've done the same experiment.

2. Spin the coin instead of flipping it. Before spinning, predict if the results will also be close to 50/50. Investigate and find out.

3. Explore other two-sided coins or objects, such as dimes, quarters, coins from other countries, or two-color counters. Are the results of tossing these coins close to 50/50? Have students make predictions about the results before they conduct the investigation.

4. Assign coin flipping investigations to be done as homework with family members. Have students report their data, on a graph or in writing.

Penny Flip

1. PREDICT how many heads and tails you will get if you flip a penny 20 times.

PREDICTION:

(------)
Heads

(------)
Tails

2. FLIP the penny 20 times and record your results below.

Heads

Tails

3. ADD your results to the class graph.   RESULTS: (_____)_(_____)
                                                                H   T

✂ -

Penny Flip

1. PREDICT how many heads and tails you will get if you flip a penny 20 times.

PREDICTION:

(------)
Heads

(------)
Tails

2. FLIP the penny 20 times and record your results below.

Heads

Tails

3. ADD your results to the class graph.   RESULTS: (_____)_(_____)
                                                                H   T

Activity 2: Track Meet

Overview

In Sessions 1 and 2 of this activity, students have fun playing the "Track Meet" game with two different spinners. Red, yellow and blue beans (or other markers) race to the finish line. Students record the winners of the races on class graphs. In Session 3, students look at samples of graphs from newspapers and magazines, then make graphs of the class data. Students compare, both orally and in writing, how the two different spinners affected the "Track Meet" game.

Spinner 1, used in Session 1, is divided into three equal parts. In Session 2, choose either Spinner 2 or 3, depending on the experience of your students. Many third grade teachers choose to use Spinner 2, while fifth and sixth grade teachers often choose Spinner 3. On both Spinners 2 and 3, blue covers half the area, while red and yellow each cover one-fourth of the area.

What You Need

For the class:
- ❏ Transparency of Spinner 1
- ❏ Transparency of Spinner 2 or 3
- ❏ Transparency of "Track Meet" game board
- ❏ 3 beans or other markers
(one each of red, yellow and blue in possible)
- ❏ Transparency for graphing
(graph paper or a blank transparency)
- ❏ Overhead projector
- ❏ 2 sheets of butcher paper or large graph paper to record class results
- ❏ Marking pens to record class results
- ❏ Graphs from newspapers and magazines, at least one per student
- ❏ paper for graphing
(graph paper or blank paper)
- ❏ writing paper

Your students might like to name their spinners.

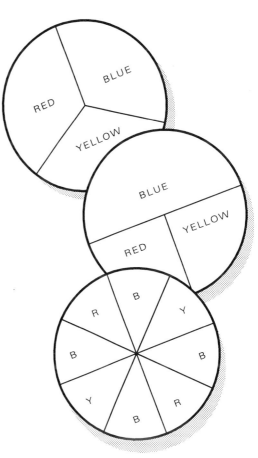

For each pair of students:
❒ Directions for Making a Spinner handout (optional)
❒ Spinner 1
❒ Spinner 2 or 3

   To make the two spinners:
   • 2 four-inch squares of cardboard or other heavy paper"
   • the spinner tops for: Spinner 1 and **either** Spinner 2 or 3
   • scissors
   • 2 paper clips
   • 4 pieces of masking tape
   • red, yellow and blue crayons
   • 2 buttons
   • ruler or other straight edge
   • pencil
❒ "Track Meet" game board (master on page 35)
❒ 3 beans or other markers
(one each of red, yellow and blue, if possible)
❒ crayons, markers or colored pencils

Getting Ready

1. Decide if your students will use Spinner 2 or 3 in Session 2.

2. Duplicate enough spinner tops so that each pair will have one of Spinner 1 and one of either Spinner 2 or 3.

3. Have students make the spinners. Each pair will need one Spinner 1 and one of either Spinner 2 or 3. Give each pair of students the materials. You can either give students the handout, "Making a Spinner" and let them work with their partner, or you can give them the directions orally. Directions for making spinners are on page 22.

4. Duplicate the "Track Meet" game board onto card stock or other heavy paper so that each pair of students will have one.

If you would like the beans to correspond to the colors on the spinners, you can spraypaint them. It is preferable to do this work outside. If you must work indoors use a very well-ventilated room. Spread beans (large white lima beans work nicely) on newspaper, making sure that the beans are not on top of each other. Follow the directions on the can of spraypaint for use and drying time. You only need to spray one side of the beans. Be sure to make plenty of extras, as they tend to peel, break and get stepped on.

One teacher modified the spinners to work better on a hot damp day. "Their sweaty little hands caused the square under the spinner to curl and warp. We scotch-taped them to large plastic lids. The kids used the eraser end of the pencil to initiate the spin."

5. Make transparencies of the "Track Meet" game board and the spinners your class will be using.

6. Prepare two class data graphs, for Sessions 1 and 2. Whether you are using butcher paper or large graph paper, cut the sheet in half lengthwise. Tape together the short ends to make a long thin strip. On both data graphs, label a column or row (depending on how you choose to orient the graph in your classroom) for each of the three colors: red, yellow and blue. Title one graph "Track Meet—Spinner 1," for Session 1, and the other "Track Meet Rematch— Spinner 2 (or 3)" for Session 2. See "Representing Data" on page 68 for more information and ideas on graphing.

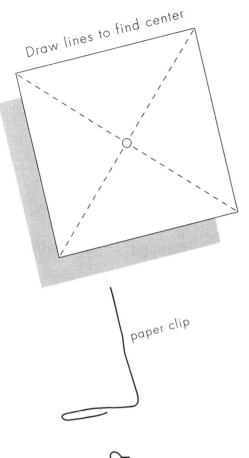

Draw lines to find center

paper clip

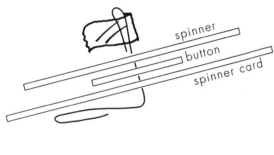

spinner
button
spinner card

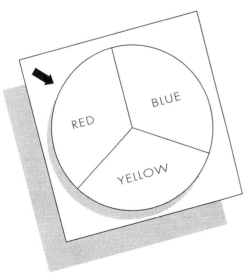

RED BLUE YELLOW

Making A Spinner

What You Need
⇨ a four-inch square of cardboard
⇨ the spinner top
⇨ scissors
⇨ a paper clip
⇨ masking tape
⇨ red, yellow and blue crayons
⇨ a button
⇨ ruler
⇨ pencil

1. Color the sections yellow, red and blue as indicated on the spinner.

2. Cut out the spinner top.

3. Draw lines diagonally across the back of the cardboard square. Where they meet is the center of the square.

4. Unfold a paper clip by pulling out the middle section and bending it upward and straightening it.

5. Poke the paper clip through the midpoint of the square.

6. Tape the paper clip to the back of the spinner.

7. To make a washer, put the button on the paper clip.

8. Put the spinner top on the paper clip, making sure you poke through the exact center of the spinner.

9. Fold the end of the paper clip down and wrap a small piece of tape around it.

10. In one corner of the square, draw a small arrow. This will be the pointer. To spin, hold the edge of the square with the fingers of one hand, and spin the spinner top with the other. Which color is the arrow pointing to?

HAPPY SPINNING!!!

Session 1: Track Meet (Spinner 1)

Students learn and play the "Track Meet" game with Spinner 1. They keep a class graph of the winners, and look for relationships between how the spinner looks and the winners of the track meets. Spinner 1 is a "fair" spinner, since each color is exactly one-third of the spinner, and each color has an **equally likely chance. It is important for you NOT to mention this at the start of the activity—allow students to think about and discuss the results in their own way first.**

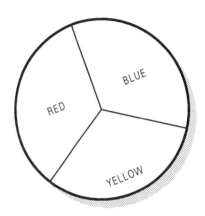

Introducing the Track Meet

1. Tell students that they will learn a new game called "Track Meet," and they'll use spinners to play. Ask students what games they already know that use spinners. Have students describe the spinners used in each game they mention.

2. Show Spinner 1 on the overhead and explain that students will use one like this today. Ask for observations about it.

3. Place the transparency of the "Track Meet" game board on the overhead. Ask students if they have ever participated in a track meet. Tell students that, in this game, three "runners," represented by beans or other markers, compete to cross the finish line first. Runners move forward according to the spinner. Two players take turns spinning. The player who spins moves whichever "runner" (colored bean) the spinner indicates one space forward. The first "runner" to cross the finish line wins.

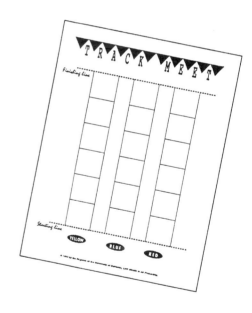

4. Survey students about the color they think will win. Ask several to explain their predictions.

5. Demonstrate the game with a student volunteer or two. Pause during play so students can look at and comment on the results of the race thus far.

6. Record the winner on the class data graph, and tell students they should record the winner after each game they play with their partners.

Discuss your expectation that everyone will accurately record the winners on the class data graph. Students occasionally want their chosen color to win so much that they falsely record results on the class data graph. You may wish to counter this tendency by initiating a discussion with the entire class. Pose questions, such as: If one pair of students reports false data, for example, recording that yellow won ten times, even though that was not what happened, what will that do to the class results? Why would someone want to make up results? What would happen if someone recorded the results inaccurately for a class election?

8. If your students would benefit from a discussion about sharing the tasks and responsibilities of working with a partner, ask for suggestions from the group and/or recommend ways to divide the work fairly.

Collecting and Reporting Data

1. Tell students to make a prediction and share it with their partner at the start of every new game.

2. Distribute Spinner 1, three beans and a "Track Meet" game board to each pair of students.

3. Allow students ample time to play several games and record the winners on the class graph.

4. Collect the materials.

5. Ask students what they noticed as they played the game. Find out if anything surprised them.

6. Direct students to look at the class data graph and ask for "true statements." Encourage students to come up with as many "true statements" as they can. If they need some guidance, answers to questions such as these can be given in "true statements": Did red or blue win more often? How many more times did yellow win than blue? How many times was blue the loser? How many races were there all together?

Be prepared for students to be noisier than usual while they are playing, due to the excitement of the game and the cheers for a winning color. Your students are likely to become very immersed in the activity and not to need your help once they get started, giving you an opportunity to circulate and watch your class to assess their levels of involvement and understanding. You are likely to find students talking and having lots of fun at the same time as they are involved and "on task."

THIRDS OF A SPINNER
To visually demonstrate to students that the size of each color on Spinner 1 is identical, you can use a transparency of the spinner that has been cut into thirds. Using an overhead projector, you can place the three pieces on top of each other to show that they are equal, and then reassemble them to show that they make a circle. Some students may want to try this themselves.

7. Put Spinner 1 on the overhead again. Ask: How much of the circle does each color cover? What fraction of this spinner is: Red? Blue? Yellow?

8. Encourage fifth and sixth grade students to notice connections between the spinner and the results on the graph. What does the spinner have to do with how the track meets came out? The red, blue and yellow areas on the spinner are equal—did each color win about an equal number of times?

Third graders and many in fourth grade may have difficulty making connections between the design of the spinner and the results of "Track Meet." For younger students, recording data, "reading" the graph, and making sense of information are the important and appropriate tasks. Deeper understanding of relationships can come later. Older students are more likely to make and articulate the connection between the class data and the spinners.

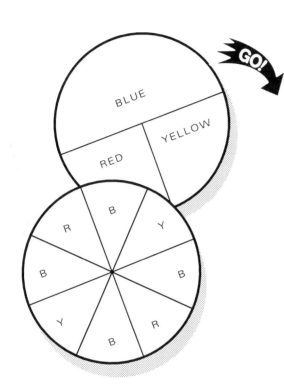

Session 2: Track Meet Rematch (Spinner 2 or 3)

Students play another series of Track Meet games, this time with another spinner. The spinner used for this session does not give equally likely chances for each color. Do not mention this to the students at the start of the activity, but listen to their comments as they compare the two spinners to each other.

Preparing for the Rematch

1. Tell students that they will conduct a rematch of the "Track Meet," using a different spinner. For review, ask for a report about what happened during the previous session.

2. On the overhead, show the spinner that they will use today (either Spinner 2 or Spinner 3) as well as Spinner 1. Ask the class what they notice about the two spinners. How are the two spinners the same? How are they different?

3. Ask how students think the track meet might be affected by using this new spinner?

4. Remind students to make a prediction and share it with their partner at the start of every game. Post the class data graph for Session 2 so the students can record the winners.

Playing the Game

1. Distribute spinners, beans and game boards.

2. Allow students time to play the game several times.

3. Collect the materials.

4. Ask students what they've noticed about playing the "Track Meet" game with the new spinner. Have students discuss the results of today's games and make comparisons with the earlier round of "Track Meet." In the next session, students will look more closely at the differences between the results of today's rematch and those from the first "Track Meet."

Session 3: Graphing and Class Conclusions

In this session, students look at graphs from newspapers and magazines to examine a variety of ways graphs are constructed. They then make graphs of the data from the two sessions of track meets.

Exploring Graphs from Newspapers and Magazines

1. Give each pair of students two or more graphs from newspapers and magazines. Ask each pair to choose one graph and find out as much as they can about.

2. After a few minutes, have two pairs join together to make groups of four. Each pair will now show the graph they chose to their two new group members and explain it.

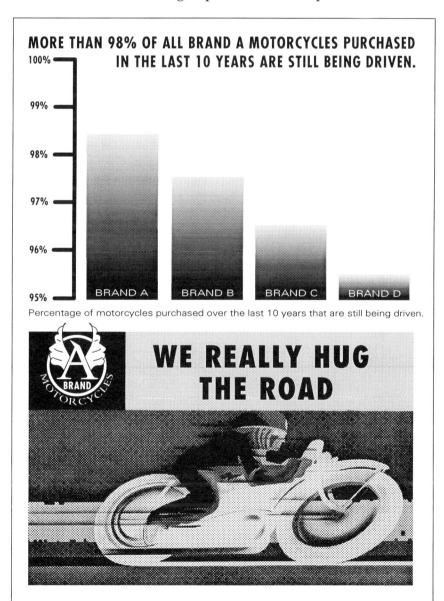

The graph demonstrates how the ways data is displayed or presented can strongly affect how we interpret it. In this case, for example, starting the graph at 95% instead of 0 allows small differences between different brands to look like much larger differences. The eye compares the bar for BRAND D against that of BRAND A, and assumes that BRAND D must be much worse. Actually, the percentage difference between them is about 3%, with BRAND D a bit over 95% and BRAND A a bit over 98%. Have your students find other examples of ways that graphic displays or other techniques can strongly influence the message or even mislead the consumer.

Some teachers use articles from the newspaper to spark discussions of probability. One teacher said, "I used an article on predicting earthquakes to talk about probability. Also, the election was a good opportunity to talk about chances." These are good real-world connections. Of course, the probability of flipping a coin or spinning a spinner can be calculated relatively easily, while earthquakes and elections both involve so many variables that they defy easy prediction.

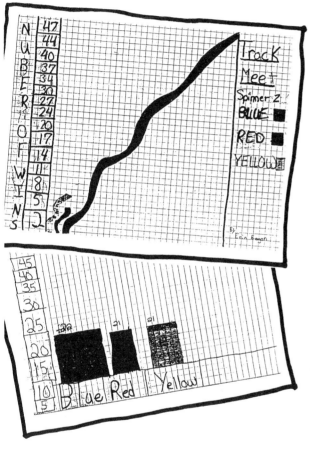

For Grade 3, you can group two or three pairs of students together and have them compile and graph the data for their group only, not for the whole class. For older students, you may want to suggest or require that they represent data from both sessions on the same graph.

3. Have groups report what they found to the whole class. Ask what kinds of things made a graph easy or difficult to read and understand. Ask students to look for titles, labels, pictures, keys or other details that they might want to include in the graphs they make. Post some of the graphs around the room for students to use as reference for the remainder of the session.

Graphing Track Meet Data

1. Post the class graphs from the two previous "Track Meet" sessions and briefly review the results with students.

2. Tell students they will work in pairs, with one making a graph for the first "Track Meet" and the other making a graph for the rematch. Let them know that their graphs should be clear and readable, interesting to look at, and convey all the important data the class gathered. Encourage students to use the posted graphs for ideas and inspiration. Students can graph on either graph paper or blank paper.

3. Circulate among the students as they work. Teachers often comment that many of the most stimulating discussions among students take place while they are graphing. Active involvement in manipulating and representing the data allows enough time for students to understand, question and analyze the results.

4. Have each pair of students make comparisons between their two graphs. Students can record what they observe by keeping two lists, one telling the ways the two graphs are similar and the other the ways they are different.

5. Post students' graphs. Ask what they noticed about the graphs. Have them share and compare their lists of similarities and differences.

6. Ask students how they might explain the differences in the results of the "Track Meet" using the two spinners. Ask what the results have to do with the spinners they used.

Comparing the Spinners

For more information about the probabilities of the spinners, see "The Probability in the Activities," page 74.

1. On the overhead, display both spinners used by your students, and ask for comparisons. How much of each circle does Blue cover? Red? Yellow? Encourage students to use fractions (and percentages, if appropriate) to describe the arrangement of the colors. If your class used Spinner 3, they may need some guided questioning to discover the fraction or percent of the circle covered by each color.

2. Discuss the idea of "fairness." Ask children what fairness means to them. Find out if they think one spinner was "more fair" than the other. Ask how they would decide if a spinner was fair or not.

3. Have students pretend they are the yellow "runner" in "Track Meet." Ask which spinner would give them a better chance of winning. Now have students imagine the are the blue "runner" and ask again which spinner would give them a better chance of winning. Ask studer.ts to explain their thinking.

4. Regarding Spinner 2 or 3, ask if one color has a greater chance (a greater probability) of coming up if you spin just one time. Ask students to explain what it means when someone says that "you have a greater *probability* of spinning blue than either red or yellow."

5. Have students recall their results in the penny flip and ask if either spinner represents an **equally likely chance**. Is there anything you noticed about Spinner #1 that is like flipping a penny? How about Spinner #2 (or #3)? Ask students to explain their reasoning.

6. Explain that mathematicians have studied the *probability* of the results that can happen from flipping coins and spinning spinners. For the "Penny Flip," mathematicians have discovered that each side of the coin has an **equally likely chance** of landing up. With an equally-divided spinner, such as Spinner 1, each color has an **equally likely chance** of being spun.

7. For older students, ask them to use fractions (and decimals if appropriate) to represent the portion of each color on the two spinners. Explain that these fractions also represent the probability of spinning the colors. Look at the class data and compare it to the fractions.

More Spinner Fractions
*Students can learn more about the relationships between the colors on the spinner by using a transparency of the spinner that has been cut into sections. For Spinner 3, you can demonstrate that Red plus Yellow equals Blue. Show that Red and Yellow are equal, and that Blue equals half of the spinner. Remind students that the data they generate from an experiment (the track meets) will not **exactly** match the fractions of the colors on the spinners, but will probably be close. In general, the larger the number of trials, the more closely the results will match the probability. This is called the **law of large numbers.** For more information on the law of large numbers, see pages 74.*

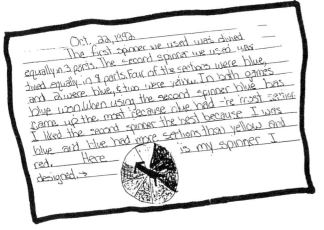

Oct. 22, 1992. The first spinner we used was divided equally in 3 parts. The second spinner we used was divided equally in 8 parts. Four of the sections were blue, 2 were yellow. In both games and 2 were blue, & two were yellow. In the second spinner blue has blue won. When using the second spinner blue has come up the most because blue had the most section. I liked the second spinner the best because I was blue and blue had more sections than yellow and red.   Here _____ is my spinner I designed. →

Oct. 22... We ve been doing a bean race. We used 2-spinners. The first spinner was divided into equal parts. The second had half blue and 1/3 of red, 1/3 of yellow. Then we made graphs on the bean race. They were double bar graphs. Some went vertical, some went horazontil. I liked All of them had more blue than red or yellow. I liked the first spinner the best because it was equal and that made the races closer. It's more exciting when they're closer together. Here's a spinner I made.

BLUE GREEN
RED YELLOW

Writing

Writing and drawings are excellent ways for students to reinforce and extend what they've learned, and can help you assess the understanding of individual students. You may want to have students share their ideas in collaborative groups before they start writing. Below are some suggestions you may want to use for writing or assessment activities.

• Write at least three "true statements" about each graph you made with your partner. More experienced students can write three questions that can be answered by looking at the graphs.

• Have students write on: Which spinner did you like using the most for Track Meet? Why?

• Draw one fair and one unfair spinner. Write about how you know each is fair or unfair.

• Use a graph from another area of study, perhaps science or social studies, or from a newspaper or magazine. Have students discuss, in small groups, what it means and then write individually about what they are able to find out from the graph.

Going Further

1. Have students conduct more trials, or compare their results with those from another class.

2. Students can design and make their own spinner, using a blank spinner, then make up a game board to go with it. They should name their game, write down at least three rules, record their predictions, play the game, record their results, and then write about what happened.

3. Involve sixth graders in middle school geometry concepts by having them use a compass and protractor to draw pie charts (circle graphs) of the data. Have students use a compass to draw a circle. They can draw a line to divide the circle in half. They can draw a line to divide one half into two fourths, and then a line to divide one fourth into two eighths, and so on. You many want to have them make another circle that they divide into thirds, sixths and so on. Students should label the fractions of the circle.

Using these divided circles as guidelines, students can estimate what fraction of the circle each color should cover. By coloring in another circle appropriately, they will have a pie chart that approximates the class results.

For a more accurate pie chart, students can calculate how many degrees of the circle should be allotted to each color, and then use a compass and a protractor to make a pie chart. For example, if red came up 20 times in 65 spins, red came up 20/65 of the time. To find the degrees, multiply the fraction (20/65) by the total number of degrees in a circle (360 degrees). Have students compare the pie charts they draw with the spinners themselves.

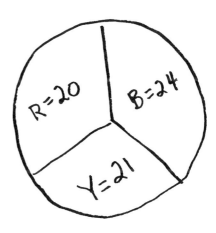

Spinner 1

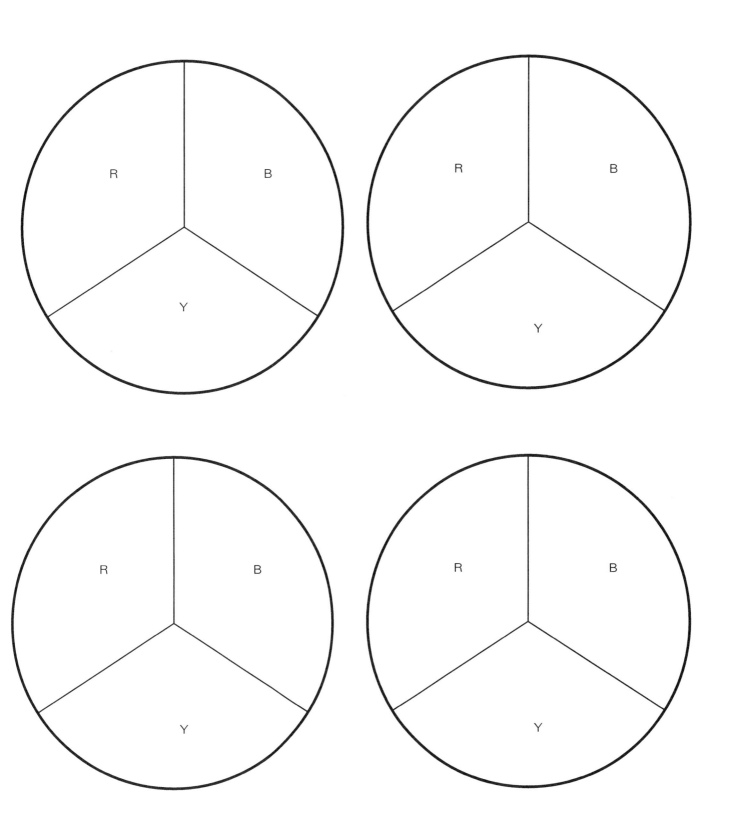

Spinner 2

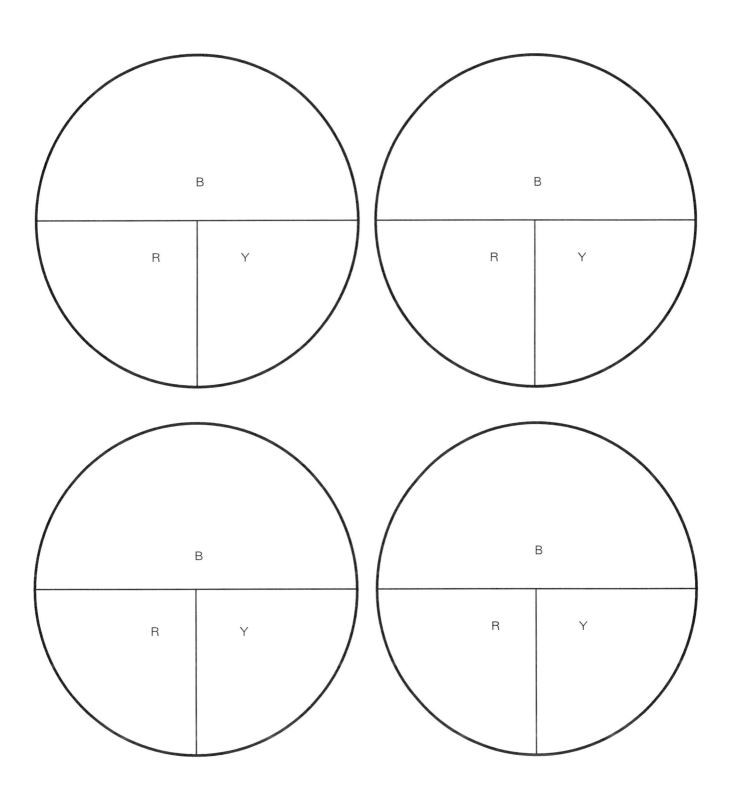

Spinner 3

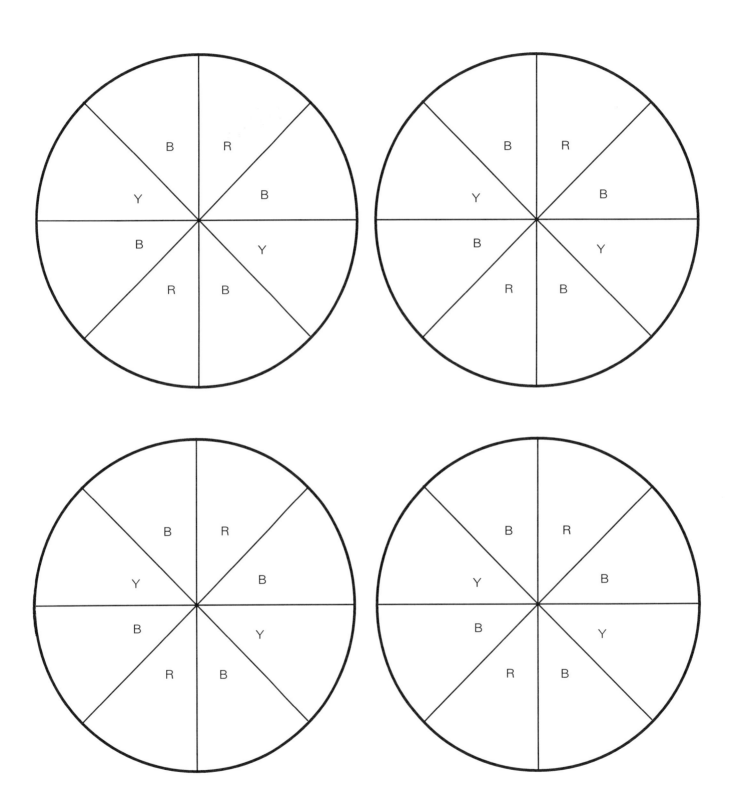

T R A C K   M E E T

Finishing line

Starting line

YELLOW   BLUE   RED

Activity 3: Roll a Die

Overview

Dice in various forms are found in many parts of the world, and have been used in both modern and ancient games of chance.

In this activity, students use standard dice: cubes with dots representing the numbers one through six. Pairs of students roll one die to generate data, then discuss and analyze the results. The experience from this activity will lay the foundation for further explorations of probability in **Activity 4: Horse Race,** a game that uses two dice.

"Die" is the singular form of the noun; "dice" is the plural form.

What You Need

For the class:
❐ 1 "Roll a Die" Data Sheet or graph paper transparency
❐ 1 die
❐ 1 overhead projector
❐ 1 overhead pen
❐ writing paper
❐ plastic berry basket (optional)

Children love things that are much bigger than usual. You can make your own giant foam dice to use for demonstrating dice activities for the whole class. Get large foam cubes (4" cubes are great) and use permanent pens to make the dots. Look at a standard die to see how the numbers are arranged: the two numbers opposite each other always add up to seven!

For each pair of students:
❐ 1 die
❐ 1 "Roll a Die" data sheet (page 42) or graph paper
❐ crayons, markers or colored pencils
❐ plastic berry basket (optional)

Getting Ready

1. Duplicate the data sheet or graph paper, for each pair of students to have one, and make one copy on a transparency.

2. For more information on graphing, see "Representing Data" on page 69.

If students shake the die and roll it into the berry basket instead of onto their desk, it is much quieter and also keeps dice from rolling onto the floor.

If students make "unreasonable" predictions, don't comment on them for now. Students' statements at this time can help you assess their level of understanding about probability. Remember, it takes much time and many experiences for anyone to develop a solid understanding of probability. Look for evidence of thinking and questioning, rather than for mastery of concepts.

Session 1: Gathering Data

In this session, students make predictions and gather data about what happens when you roll one die many times.

Rolling a Die

1. Ask students to share what they know about dice. Ask what games they have played that use dice, and why they think dice are used.

2. Tell students that in this activity they will use dice for more investigations into probability. When you roll a standard die, what are the possible numbers you can get? Ask if one number will come up much more frequently or if is there an equally likely chance of getting each number.

3. Ask students to take a moment to think about how many of each number they predict will come up in 30 rolls. Ask students to explain their predictions. Help them use what they learned with coins and spinners to predict the probabilities with dice. Do all six numbers have the same chance? Would you say that your prediction is more like a wild guess or an educated guess? You may want to have students record their predictions to later compare to their results.

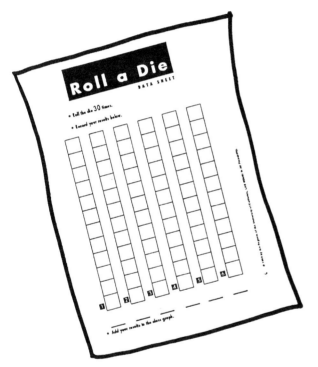

4. Put the transparency for the data sheet on the overhead. If students will make their data sheets, use a blank transparency or a graph paper transparency. Otherwise, use the "Roll a Die" data sheet.

5. Have a volunteer roll a die 30 times while you record the results on the overhead projector. Stop several times during data collection to allow time for students to analyze the results and comment on the data.

Some teachers found that completing 30 rolls during the demonstration was too much. Again, you are in a position to make the best decision for your class regarding how much modeling they need, and/or how much they enjoy a whole group activity.

6. Ask for "true statements" about the data. Ask students to predict the results for the whole class.

Collecting and Reporting the Data

1. Distribute a "Roll a Die" data sheet graph paper or blank paper to each pair of students. Allow time for students to make their own data sheet, as needed. Students should make a prediction for how many times they think they will roll each number. Remind students to keep track of their rolls.

2. Distribute one die to each pair of students. Have students roll the die and chart their data.

3. Have a few students report their data and observations to the class. What happened? Were your predictions close to your results? What surprised you about the "Roll a Die" investigation?

4. Encourage discussion about the experiment. If appropriate for your group, ask, "Do you think a die is fair or unfair?" Recall the two spinners: is the die more similar to Spinner 1 (equal divisions) or to Spinner 2 or 3 (unequal divisions)?

5. Save student data sheets for the next session.

Session 2: Graphing and Class Conclusions

In this session, students graph class data and then discuss what they have found.

GO! ➤ **Graphing the Class Data**

1. Hold a class discussion to briefly review the results from "Roll a Die."

2. Post the numerical results from each pair so they are accessible to all students.

3. Students should calculate the total for each number (how many ones did the class roll? etc.) and report their answers. When students disagree on the total, have everyone recalculate until all agree on the answer.

4. Ask students to recall what should be included in their graphs. (Title, labels, etc.)

For younger students, group two or three pairs together, then let them calculate the totals and graph the results.

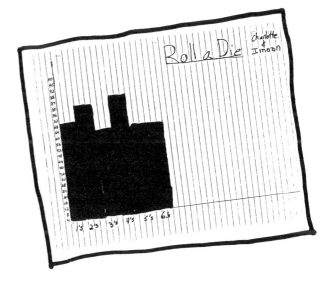

One teacher commented: "Designing graphs for the data from the whole class, especially when each square had to count for multiple tosses or flips, was a very good thinking and planning activity."

With one die, on each roll, there is one chance of getting a 1, one chance of getting a 2, one chance of getting a 3, etc. However, when two dice are used, as they will be in the next activity, the odds (or chances) change. For example, with two dice, there is not the same chance of getting a 2 as there is of getting a 7.

5. Have students construct their graphs.

6. Post the students' graphs and ask students to explain the results.

7. Ask students what connection they see between the dice and the data on the graph.

8. Show older students numerical ways to represent probability when rolling one die. For example, there is a one in six chance of getting a one on every roll. The probability of getting a one is 1/6. There is a five in six chance of getting a number less than six on every roll; the probability of getting a number less than six is 5/6. See "Probability in the Activities," page 74 for more information.

As with any probability activities, the data generated and graphed by your students may or may not closely match the actual probability. In this activity, the results from some classes do show that each of the numbers 1 through 6 occurs roughly an equal number of times. However, results from other classes yield data that is heavily weighted in one direction or another. **This kind of variation in probability data is normal.** It does not mean the dice are faulty. For more information, see "Is Almost Good Enough?" on page 75 and "The Law of Large Numbers" on pages 74 and 75.

Writing

Many teachers have found that when students have an audience for their writing, they are more motivated, more creative, and include more details in their written work.

- Have students write a letter to a parent, grandparent, principal, janitor or other adult to explain everything the student knows about dice, including information from the experiment they have just conducted. Students may enjoy sharing the letter with the addressee.

- Have students write a letter to a company that makes dice and explain an idea for a new kind of die where the number six comes up more often than any other number. Students can illustrate their letter.

Going Further

1. Have students consider this situation: You and your sister flip a penny to see who will take out the garbage. The penny rolls under the refrigerator, and neither of you can get it out. There are no more coins in the house, but you have one die. How can you use the die so that both you and your sister have an equal chance of taking out (or not taking out!) the garbage?

2. With older students, have them compare the results of a small number of rolls (for example 10 or 30 rolls) with those of the whole class. Ask them to describe what they notice, and then offer some possible explanations.

Explain that mathematicians who study probability usually collect a lot of data. They would want to collect data from **many** flips of a coin or rolls of a die. The **Law of Large Numbers** says that a small number of trials of dice tosses or coin flips will yield a wide range of results, which may or may not be close to the expected probability. However, a large number of trials, tends to give results that are quite close to the expected probability. For example, if you toss a die six times, it is not very likely that you will get one of each number; if you toss a die six hundred times, you will almost certainly get *about* 100 of each number.

Non-standard dice, with different markings or with more or less than six sides, are available from teacher supply and game stores.

Students will probably notice this phenomena in the activities of this unit: their own results (a small number of trials) may not be close to the expected probability, while the class results (a larger number of trials) is likely to be quite close.

For more information, see "the Law of Large Numbers," pages 74.

3. Use non-standard dice to conduct further "Roll a Die" experiments. Use dice that are four-sided (tetrahedra), twelve-sided (dodecahedra), or other unusual shapes. Before conducting the trials, students should make predictions about what they think will happen. Graph the results.

Roll a Die

- Roll the die 30 times.

- Record your results below.

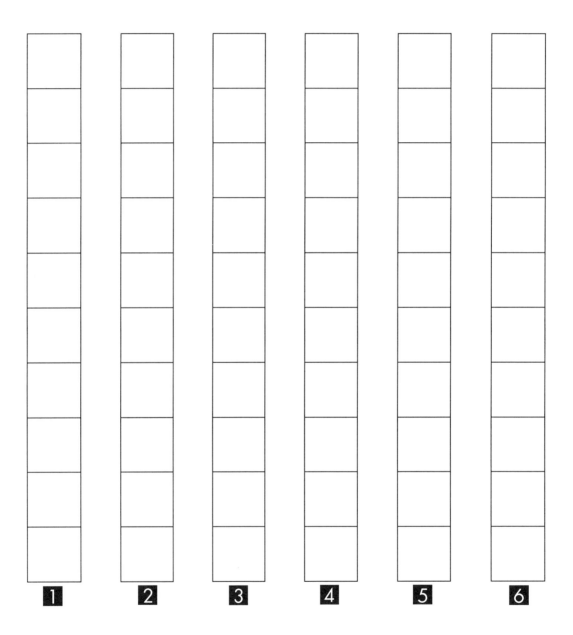

1   2   3   4   5   6

TOTAL —— —— —— —— —— ——

- Add your results to the class graph.

Activity 4: Horse Race

Overview

In this activity, students play "Horse Race," in which twelve "horses" compete to cross the finish line first. The horses are numbered one through twelve. Two dice are used. A particular horse moves towards the finish line when the sum of the two dice rolled equals its number. For example, Horse 5 moves ahead one space if a four and a one are rolled. Students make predictions about which horse they think will finish first. Cheering on their favorite horses, students play the game many times to generate data.

After the game, students grapple with ideas about why the horses do not win roughly an equal number of times. This activity presents a captivating way for students to take a spirited interest in the outcome of a probability experiment.

What You Need

For the class:
- ❏ 1 "Horse Race" game board transparency
- ❏ 12 beans or other small counters
(large lima beans, cubes, or chips work well as counters.)
- ❏ 1 "Keeping Track" transparency
- ❏ 1 overhead projector
- ❏ 2 overhead pens, if possible in the colors that match the dice students will use in the second session of this activity
- ❏ 1–2 sheets of butcher paper or large graph paper
- ❏ writing paper

For each team of 2 students:
- ❏ 2 dice, each a different color
- ❏ 1 "Horse Race" game board (page 52)
- ❏ 12 beans or other small counters
- ❏ 1 plastic bag or other small container for the counters
- ❏ 1 "Keeping Track" chart (page 51)
- ❏ crayons to match the colors of the dice

Note: A larger version of the game board is provided at the end of the book.

Getting Ready

1. Duplicate a copy of the "Horse Race" game board for each pair of students, and make one copy on a transparency.

2. Duplicate a copy of "Keeping Track" chart for each pair of students and make one copy on a transparency.

3. Make a class graph on a large piece of butcher paper or graph paper. Label one axis with numbers 1–12 for each of the horses. Label the other axis "number of wins." Title the graph. (You might also decide to have two or three students make the class graph.)

Session 1: Racing the Horses

Students play "Horse Race," using two dice, and keep track of the winning horses.

A Day at the Races

1. Have students review what they found out in the last activity. If students wrote letters at the end of the last activity, have a few read theirs aloud as the review. You can ask students what information or ideas the writer included, and if there is anything else that could be added.

2. Tell students that today they will play a game using two dice. Place the "Horse Race" game board on the overhead and put 12 beans on the starting line, one to represent each horse.

3. Explain how the horses move across the track. After a player rolls the dice, he moves the horse **whose number is the sum of the dice** ahead **ONE** space. For example, if a six and a three are rolled, the player moves Horse #9 **one space** forward.

4. Have students predict which horse will reach the finish line first. Take several responses. Have all students predict, by a show of hands, which horse they think will win.

5. Demonstrate the game on the overhead with the transparency of the "Horse Race" game board. Select two students to model taking turns rolling the dice and moving the appropriate horses.

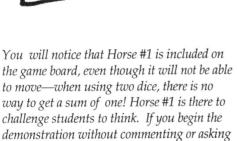

You will notice that Horse #1 is included on the game board, even though it will not be able to move—when using two dice, there is no way to get a sum of one! Horse #1 is there to challenge students to think. If you begin the demonstration without commenting or asking about Horse #1, a student is very likely to point this fact out soon, and then the class can discuss it.

6. Pause frequently to look at the results. Ask students if they notice any patterns in the way the horses are moving. Ask students to comment on why they think some of the horses haven't moved yet. Students will want to predict again as the race progresses, changing their favorite as they watch the race. Remember, it's okay to change your prediction as you receive new information.

7. Post the class graph and record the winning horse.

Playing the Game

1. Remind students about accurate recording of their results.

2. Distribute "Horse Race" game boards, counters, and dice.

3. Have students begin playing "Horse Race." Each race should be recorded on the class graph as soon as it is over.

4. Provide time for discussion of thoughts on the "Horse Race" game. Ask students what surprised them about the results. Ask for "true statements" about the graph. Save the graph for Session 2.

Play It Again

If possible, arrange for time that students can play "Horse Race" again before doing Session 2. Additional winners should be added to the class graph, so there will be more data to analyze in the next session. Additionally, students can informally test out their own hypotheses about which horses usually win, and continue to develop a broad base of experiences that will lead to a better intuitive as well as mathematical understanding of probability.

Session 2: How Many Combinations?

In this session, students complete a chart showing all the possible ways that two dice can land, and use this information to help explain the surprising results of the Horse Race.

Playing a complete game of "Horse Race" with the whole class generates a great deal of enthusiasm among the students, and provides motivation that can carry through to their play with partners. One teacher commented that, "Demonstration games seem helpful in that they give average to low students more direction as to how to proceed, and everyone seems to enjoy them."

Sometimes students want their favorite horse to win so much that they will, if not monitored, record any number of unearned "victories" on the class graph. If you anticipate this as a problem for your students, you can initiate a discussion on the need for accuracy in reporting data. Why does this matter? Discuss the situation of giving horses an "unfair advantage." What would happen if scientists doing experiments falsified their data?

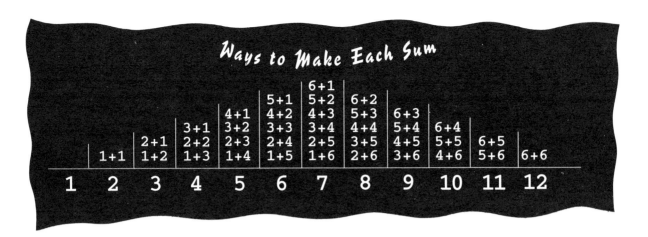

Ways to Make Each Sum

1	2	3	4	5	6	7	8	9	10	11	12
						6+1					
					5+1	5+2	6+2				
			4+1	4+1	4+2	4+3	5+3	6+3			
		3+1	3+2	3+3	3+3	3+4	4+4	5+4	6+4		
	2+1	2+2	2+3	2+4	2+4	2+5	3+5	4+5	5+5	6+5	
1+1	1+2	1+3	1+4	1+5	1+6	1+6	2+6	3+6	4+6	5+6	6+6

One teacher recorded the ways to make the sums on a chart like the one in the above illustration. When she was done, the students could see a visual relationship between the ways to make the sums and the data from the "Horse Race" game.

Ways to Make Seven with Two Dice

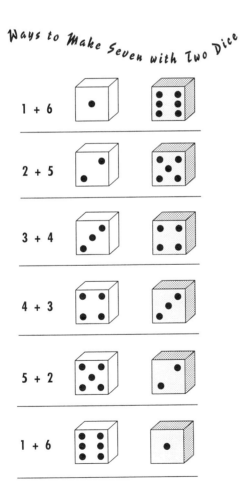

1 + 6

2 + 5

3 + 4

4 + 3

5 + 2

1 + 6

All Possible Ways

1. Post the class graph from Session 1 and briefly review the results. Ask students if they can think of any explanations for the results shown on the graph.

2. To help students look more closely at the mathematics involved in "Horse Race," ask, "How many ways could you make a seven with your two dice?" Tell them this can include reversals, such as 3+4 and 4+3. For example, a red three and a white four is a different arrangement of the dice than a white three and a red four. On the board, record all the possible ways to make seven using two dice.

3. List the ways to make several other numbers such as the number that won the most, or numbers that won few or no races.

"Keeping Track" Chart

1. On the overhead, show the "Keeping Track" chart.

2. To help students understand the "Keeping Track" chart, color the dice on the chart transparency. For example, if students used a green die and a red die, shade the dice in the top row green, and the dice in the vertical column red. Demonstrate how to fill in part of the chart by adding the numbers on the dice and placing their sum in the appropriate box.

3. Distribute the "Keeping Track" charts to students. Have them color the dice on the chart to match the ones they used when they played "Horse Race." Then they should fill in the sums on the chart.

4. Ask students to see **how many ways** there are to make the sums of each number from 2 to 12. For, example: there is one way to make 2; there are two ways to make 3, etc.

5. When they are done, ask students if they notice any patterns. Ask if this chart shows all the ways to make the sum seven, as well as other sums. Ask students to notice which sums can be made many ways, and which can only be made in one or two ways. Ask how this relates to the winners of the "Horse Race" game.

6. Older students can use the completed "Keeping Track" chart to figure out fractional or decimal probabilities. There are 36 possible ways (*permutations*) for the dice to land when you roll two dice. For example, the chance of getting a two is the chance of getting a seven is one out of 36 (1/36) six rolls out of 36 (6/36 or 1/6).

Writing

Writing activities can ask students to explain their mathematical thinking. Below are some examples that give students a chance to express their reasoning.

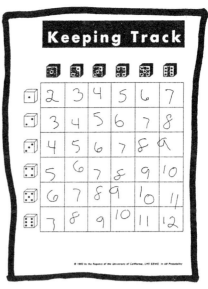

The "Keeping Track" chart can help students visualize that a green three added to a red four is a different way to obtain the sum of seven than a red three and a green four. When playing the "Horse Race" game, the sum itself is important, not the way the sum is made. When playing the game, it is the **combinations** that matter. However, when completing the "Keeping Track" chart, **the way the sums are made is important**, and these are called **permutations**.

Horse Race

1- doesn't have any chase of winning because its in posicle to roll a one with two dice

2- Theres only one chance you can roll a 2 and thats why it doesn't win. 🎲🎲 — 🎲🎲 ⬜⬜

3- Wins more than one or two but thats because it has 2 chances

4-Wins more because they have 3 chances of winning

10-Wins ūust as much as four because it has 3 chances of winning to

11- Wins less than ten because it only has 2 chances of winning

12- only has one chance of winning

• Write about the horses that are the least likely to win, and explain how you know this.

• Write about the horses that have *some* chance of winning, although they do not win often. Explain how you know this.

Going Further

1. Have students work in pairs to devise new rules for "Horse Race" so that Horse #1 has a chance of winning.

2. Obtain non-standard dice with other numbers or symbols, or with four, eight or more sides from a teacher supply or game store. Have students brainstorm investigations they could do with these dice, and predict the results. Let the class decide which investigations to try.

3. Have students bring in games that use dice and allow time for students to play them with each other. Discuss how **probability** enters into the games. Is every chance an equal one? Are there rules that allow players to get extra points or another turn if they roll an unlikely combination of dice?

4. Have students, working in pairs or small groups, make up their own games with dice. Each game should have a name and at least three written directions. Students should explain how probability is used in their game.

The GEMS teacher's guide QUADICE for Grades 4–8 features an original dice game that involves students in using mathematics skills, including exploration of probability. A cooperative version of the game is included. Playing cooperative QUADICE would make an excellent extension to these In All Probability activities.

The horses that won the least times are horse one and horse 12. I know this because when we spun the dice and added the numbers together we found that there is no way that one can be rolled, for there is no zero on any of the dice. I know that twelve can only be rolled once because six plus six is the possible combonation to make twelve.

Keeping Track

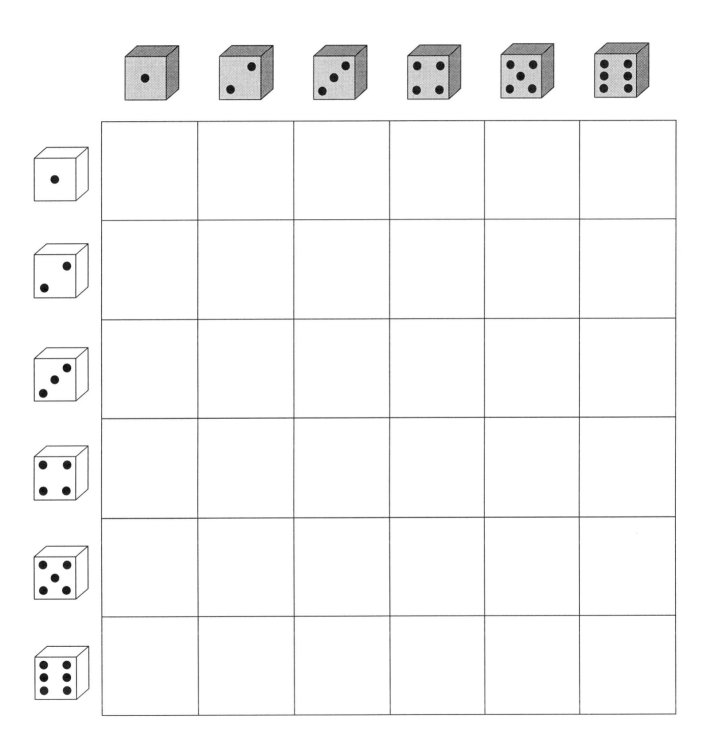

Horse Race!

GAME BOARD

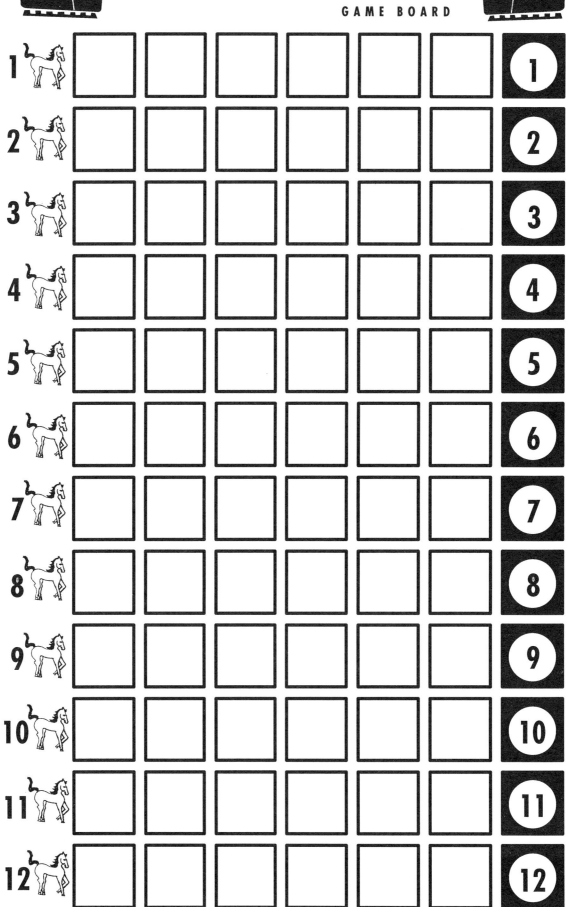

Activity 5: Native American Game Sticks

Overview

This game is based on a Native American game of chance. Each person (or team) uses six two-sided sticks. Counters are awarded based on the number of sticks that land with the design side and the plain side facing up. Originally, the sticks used for this game were carved with geometric decorations. For this activity, students make their own sticks by drawing designs on wooden tongue depressors.

Native American Game Sticks involves intricate probabilities, and most students will not analyze the game in the same detail as the previous games and activities. Given students' prior experience with coins, spinners and dice, they are likely to gain an intuitive sense that probability is also involved in this complex and enjoyable game of chance. The idea here is to introduce students to a game of chance that they most likely have not played before, to stimulate their growing awareness of probability, and to suggest further questions for exploration.

See "Native American Game Sticks," page 81, for more information on Native American games of chance. For more information on the probability involved in this game, see "The Probability in the Activities" on page 74.

What You Need

For the class:
- ❐ 12 Game Sticks (see "Getting Ready" below)
- ❐ 10 counters (toothpicks or cubes, for example)
- ❐ 1 Traditional Designs from California transparency
- ❐ 1 overhead projector
- ❐ writing paper
- ❐ blank overhead transparency, optional
- ❐ butcher paper, optional

For each team of two students:
- ❐ 12 tongue depressors (3/4" or 1" wide) available from teacher supply, craft, and drug stores
- ❐ colored marking pens
- ❐ 1 "Traditional Designs from California" handout
- ❐ 1 container of 10 counters (beans, chips, cubes, or other small items)
- ❐ graph paper (optional)

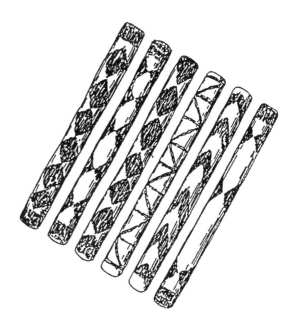

Younger students can understand that certain combinations of sticks are more common than others, and that the more common combinations have a greater probability of occurring. In Session 2 of this activity, older students can explore simpler versions of Game Sticks in order gain numerical insight into the complex probabilities.

Sometimes teachers feel they must "know it all" before they can teach. One teacher countered that notion this way: "I think it is important for my students to see that I do not know everything, and to be reminded that even smart, educated grown-ups don't know it all. Nobody does. I like to model to my students that it's okay to admit I don't know everything, and that I can learn something new, too, even at my age. It's a valuable lesson for them."

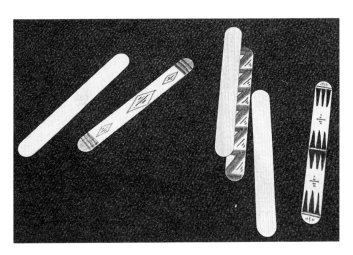

Getting Ready

1. Read over the information on "Native American Game Sticks," page 78.

2. Make a copy of "Traditional Designs from California" for each pair of students and make one copy on a transparency.

3. Use colored marking pens to make two sets of game sticks by decorating 12 tongue depressors (six sticks per set). Remember to decorate one side only, and to label the other side with your name. These will be used by the students during the demonstration of Game Sticks.

4. If you want to post the rules of the game, write them on an overhead transparency or on butcher paper. See "Rules for Game Sticks" on page 57.

Session 1: Making and Playing Game Sticks

To demonstrate the game, the class plays together, with half the students on each team. Individuals then decorate their own game sticks and play the game, either in pairs or small groups. As it takes about a class period for most students to decorate their sticks, you will need to allow other time for students to play the game with their own sticks.

 Demonstrating the Game

1. Tell students that they will learn to play a game similar to one played by Native Americans in many parts of what is now the United States, Canada and Mexico. Hold up the game sticks so the class can see that sticks are decorated with a design on one side and are plain, with no design, on the other. Let students know that after they learn the game as a whole class, each student will decorate their own sticks, and then they can play in pairs or in small teams.

2. Explain that each team wins counters based on how the sticks land. Ask students to describe some ways the sticks can land.

3. Post the rules if you have prepared them in advance. Explain the rules for getting counters as follows:

Rules for Game Sticks

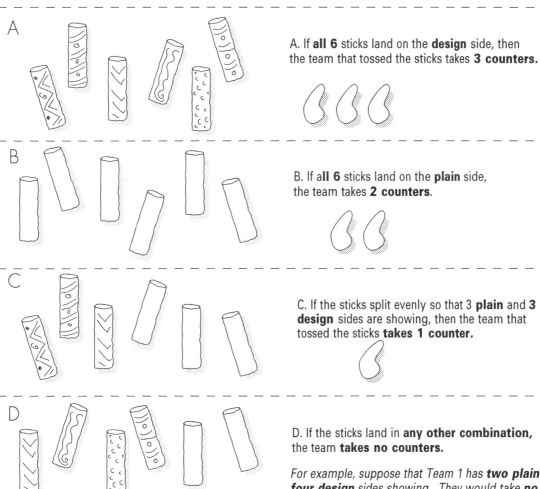

A. If **all 6** sticks land on the **design** side, then the team that tossed the sticks takes **3 counters.**

B. If **all 6** sticks land on the **plain** side, the team takes **2 counters**.

C. If the sticks split evenly so that 3 **plain** and **3 design** sides are showing, then the team that tossed the sticks **takes 1 counter.**

D. If the sticks land in **any other combination,** the team **takes no counters.**

*For example, suppose that Team 1 has **two plain** and **four design** sides showing. They would take **no counters**. Then suppose Team 2's toss shows **three plain** and **three design** sides showing. Team 2 would take **one counter**.*

E. When no counters remain in the middle, teams take the counters **from each other** when they toss a winning combination of sticks.

F. The game ends when one team has **all** the counters.

*You may want to stress the difference between tossing the sticks and throwing the sticks. Sticks should land **gently** in front of the tosser. Some Native Americans use baskets to shake the sticks, which are then tossed from the basket to the ground.*

4. Divide the class into two groups. Put 10 counters in a central location between the teams. Give one person in each team a set of six Game Sticks. Tell students that the set of sticks will rotate in an orderly fashion from student to student, so that everyone will have a chance to toss the sticks for their team.

5. To find out which team will begin, have one member from each team toss their six sticks. Whichever team has the most design sides facing up goes first.

6. A student on the first team tosses the sticks, and takes counters if indicated by the way the sticks land. Teams alternate, with a different student tossing each time.

7. Native Americans often use songs or chants to encourage their team. Your students may wish to sing or chant encouraging words to the tosser on their team. **Make it clear to students that only positive cheers for their team are acceptable, and that no jeers or other negative comments are allowed.**

8. Ask students for their comments about the game.

Making Game Sticks

1. On the overhead, display the transparency of "Traditional Designs from California," and ask students what they notice about the designs. They may mention zig-zags, repeating patterns, diamonds, squares, triangles, dots, or other design elements. Let students know that they can use these designs or come up with their own.

2. Remind students to decorate **only one side** of the stick with designs. The student's name (or initials) should be the only thing written on the other side. Names or initials should be written in small plain letters, so they cannot be confused with the design side of the stick.

3. To make a set of game sticks, each student needs six 3/4"-wide tongue depressors. Give a copy of "Traditional Designs from California" and colored markers to each pair of students.

4. Have students prepare their game sticks.

5. Students who finish decorating their sticks early can play with others who are also done.

Play the Game

1. In the time remaining, have all students, in pairs or small groups, play the game.

2. Circulate as needed, to answer questions and make sure all students are becoming familiar with the rules of the game, tossing the sticks gently, and playing cooperatively.

Play It Again

1. Provide another class session when students can play the game several times, either in pairs or small groups.

2. Afterwards, ask students for their comments and observations about the game. Ask what they noticed about the way the sticks fell as they played.

3. Refer students back to the rules, and ask why they think the inventors of the game made the rules this way.

Session 2: A Probability Experiment

In this session, older students discuss connections they find between the Native American Game Sticks game and their ideas about the concept of probability. Suggestions for further exploration are given, to be used as you determine most appropriate given your students' interest and skills. More detailed information about the probability involved in the "Game Sticks" activity can be found in "The Probability in the Activities" on page 74.

How Likely?

1. Ask students to recall their observations about how the sticks tended to fall.

2. Ask students to further describe all the possible ways the sticks could land. Draw them on the board or overhead. You may want to draw the possibilities in a logical sequence, even if students do not suggest them in this order, as a model of one way to organize data.

Six game sticks can land in these combinations

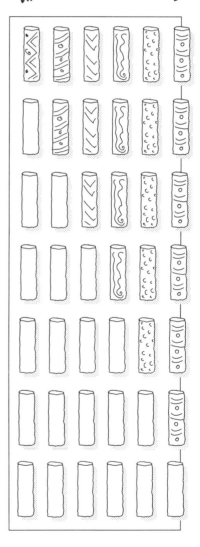

Depending on your class, you can use all, some, or none of this section, or modify/adapt it to suit your students.

3. Pose questions, such as: Are some ways more likely than others? Do you think you had more tosses when you took counters or when you didn't take counters?

4. Ask students what *probability* has to do with "Game Sticks." Ask for their ideas about how they could find out more about which combinations of sticks are most likely.

The Experiment

1. One way to investigate the probabilities involved in Native American Game Sticks is to look at simpler versions of the game. Ask students how many ways just **one** game stick can land. On the board or overhead, draw the two possibilities: design side up, and plain side up.

2. Ask students to relate tossing one stick to flipping the penny at the beginning of the unit. How are they similar? How is the probability similar

How one stick can land   How one penny can land

3. Now ask students to consider tossing **two** game sticks. Students are likely to say there are three possibilities: two design sides, two plain sides, and one of each. Actually, there are four ways the sticks can land when using two sticks. Remind them of the "Keeping Track" chart they used for the two dice: a green three and a red four is different than a red three and a green four. Similarly with the sticks, there are two ways to get one design and one plain. Draw the four possibilities on the board or overhead.

4. Ask students to work in pairs and record the possibilities for one stick, two sticks and four sticks. If students think they have all the possibilities for four sticks, ask them how they know they have them all.

How two sticks can land

(A A)

(A B)

(B A)

(B B)

The ways four sticks can land

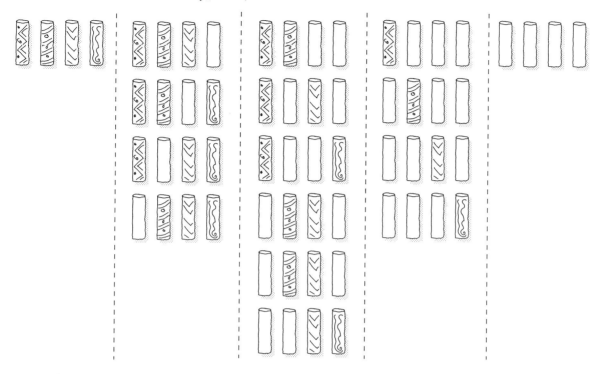

5. Ask students what they discovered about the way four sticks can land. Do they see any pattern beginning to develop? Encourage students to describe the **probability** of getting different combinations. For example, how many ways can you get all four stick to land design side up? How many ways can you get two designs and two plain?

6. Ask students what they predict would result if they were to draw **all the possibilities for six sticks**. Take several responses. You may want to ask how they think the inventors of this game incorporated a knowledge of probability when they made up the rules. How is probability incorporated into the rules?

7. Have students predict what results they think they would see if they collected data about tossing six sticks. Have each pair toss their sticks a number of times and keep track of how many design sides are up each time. Collect all the class data and have students graph it.

8. Ask students to describe the graph and give an explanation for why they got this data.

Students can make a "tree diagram" showing the possibilities when four sticks are tossed.

It is not at all necessary to arrive at a final "answer" to the six-stick question. Depending on the level of your students and your time considerations, you may want to: suggest further investigation of the six-stick probability as a special project; or just explore and discuss it sufficiently so students become aware that it is quite complex. (For your information, there are actually 64 different possibilities! See page 78-80 for more information.)

Writing

As with the earlier activities, one excellent means of extending and assessing student learning in a mathematics unit is to have students write. Here are a few suggestions relating to this activity.

• Ask the class to brainstorm words that relate to probability and this unit. Record them on the board. Have students write a letter to you or the principal and tell everything they know about probability.

• Have students answer this question in writing: "What does probability have to do with Native American Game Sticks?"

• Ask students to describe another game they like to play that involves probability and to explain how probability is used in that game.

Going Further

1. Organize a special visit to another classroom and have your students teach this game to students in a younger grade.

2. Students can use two different coins (such as a penny and a dime) to investigate the fairness of another game, "Two Coin Flip." The rules are as follows: There are two players and each gets one coin. They both flip at the same time. Player A gets a point when both coins are the same (two heads or two tails). Player B gets a point when the coins are different (one head and one tail). Points are recorded until one player gets 10 points. Who has a better chance of winning—Player A or B? Or, do both players have the same chance? Play and record results. Is this a fair game? Why or why not? Students can create a tree diagram to show the possibilities for Two Coin Flip.

3. There are many variations of Native American game sticks as well as a wealth of other games played by different Native American nations and tribes. A good resource for ideas is *Games of the Native American Indians* by Stewart Culin. You and your students can find out more about other games of chance played by Native Americans in your region of the country.

4. Students can share games of chance from their own and other cultures. They may already know a game, or they may be able to find out from a family member or friend. Students can also do research in the library to learn about other games. Play games from other cultures and discuss how they compare to Native American Game Sticks, Track Meet and Horse Race.

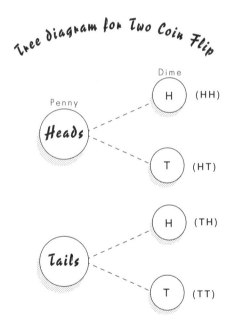

Tree diagram for Two Coin Flip

Penny

Dime

Heads

H (HH)

T (HT)

Tails

H (TH)

T (TT)

TRADITIONAL DESIGNS from CALIFORNIA

Native American Game Sticks

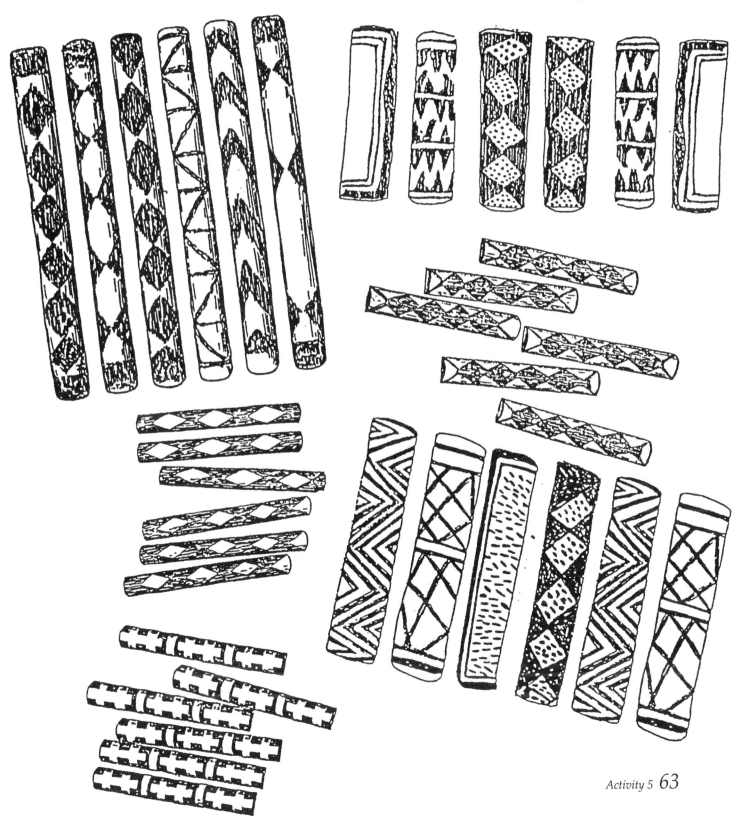

Going Further *(for the entire unit)*

In addition to the "Going Further" activities listed for each activity, there are many other possible extensions to this unit. Here are just a few suggestions:

1. Students can make up their own probability games, using sticks, dice, spinners or coins. Each game should have a title and at least three rules. Students can design the game board to go with their game.

2. Students could research and report on an occupation or profession that makes use of mathematical knowledge related to statistics and/or probability. They could interview a person in such a field. Some careers are listed in the "Behind the Scenes" section.

3. Have students investigate and report on how statistics and probabity are used in TV news reports, the front page of the newspaper, the weather page, or the sports page.

4. In addition to the writing assignments noted after each of the acivities, consider also other creative writing ideas, such as writing a short story that hinges around the likelihood that some thing or things will or will not happen. This story could be entitled "A Likely Story."

5. Students could do an oral presentation of the poem about probability on the back cover of this guide. Students could also use their experiences with this unit as a jumping off place for their own poetic recitals, in "rap" or other modern styles, set to music, or as a group dramatic reading or skit that focuses on probability or one or more of the games they played.

Behind the Scenes

Note: This section is designed to assist you in interweaving background knowledge into the activities as appropriate for your class and to provide some helpful suggestions regarding graphing and representing data. **It is not meant to be read out loud to or duplicated for students.** As in all GEMS activities, the activities are primary, and are designed to help students discover the concepts for themselves. Students always have many questions and creative ideas, and this background information should assist you in helping students explore probability in greater depth.

Probability

The probable is what usually happens.
— Aristotle

Aristotle's quotation, written hundreds of years before we know of any recorded formal mathematics of probability, is one way to think about probability without using numbers. This is a good intuitive definition, and one that you may want to share with your students.

Probability is the area of mathematics that tries to figure out and make predictions about the chance or likelihood that something will (or will not) happen. Statistics is the collection, classification, analysis, and interpretation of numerical facts and data. As students learn in this unit, the collection and analysis of data can provide much information about the probability of something happening.

Students' (and our own!) understanding of probability evolves and develops over many years and in a variety of contexts. The activities of *In All Probability* provide concrete experiences upon which students can build an intuitive understanding of probability. Some older children will begin developing a numerical understanding of probability as well. Early experiences provide students with a foundation from which future numerical and mathematical work with probability and statistics can grow. In turn, students' increased knowledge can equip them to evaluate issues and make societal decisions in a more informed way.

For all of us, the concepts of probability take time and experience to develop. If you allow students to grapple with the data, make conjectures, and investigate further on their own, they will come to understandings **at their own level**. Some students, for example, may still believe that the results of the flip of a coin or the toss of a die are governed by luck, and that there is no particular pattern to how they fall. Developmentally, this is absolutely appropriate. Nevertheless, students can still search for patterns and begin to make sense of them. Students will gain important mathematical experience with graphing and statistics, as well as probability—all valuable tools for further investigations and growth as they get older.

Many teachers have found that they learn a lot about probability while teaching their students. Don't worry that you don't know it all yet—plunge in and learn along with your students!

Wild Guesses and Educated Predictions

The word prediction combines two Latin roots. *"Pre"* means **before** and *"diction"* means **to say**. So prediction literally means "to say before."

A prediction is a guess; there is no right or wrong answer, although some predictions are closer than others. Predicting can be a fun and interesting challenge. Students will be more comfortable making predictions if their guesses are accepted unconditionally. **The goal for students is to think ahead about what might happen, not to worry about being "wrong."**

It is also important to discourage competitive aspects that may arise, as when students view a close prediction as "winning" or as "better" than another student's estimate. In order to minimize the competitive aspects, one teacher suggested encouraging students to say "I think I'm getting better at predicting," rather than, "I won," when making an accurate prediction.

A prediction made without much information can be thought of as a "wild guess." Trying to predict how many grains of sand are on the beach would probably fall into this category. A more educated prediction could be made about which horse is most likely to win the "Horse Race" in Activity 4, after many games have been played and the data analyzed.

If your students are not used to making predictions, you may want to give them some practice with **prediction jars** (also called estimation jars). Fill a jar or clear plastic container with high-interest items such as marbles, nuts, plastic animals, or polished rocks. Have students examine the jar and predict how many items are inside.

This activity can be repeated many times using different containers and items. At first, student predictions may be more like wild guesses. With experience, students will be able to make predictions within a realistic range. Students often enjoy relating their method of making a prediction. By listening to methods used by other students, everyone can gain other ideas that can be helpful. Predicting, like any skill, improves with practice.

Representing Data

Graphs are one of the most powerful tools in statistics. Graphs are used in mathematics, science, social studies and other fields as visual means to represent numerical data. The ability to construct and interpret graphs is vital to mathematical literacy, and is a great aid to thinking and communicating in general.

The activities in *In All Probability* require students to use data sheets and make graphs. **Data sheets** are used for recording data generated by a pair of students, for example, what results they get when they flip a penny 20 times or toss a die 30 times. A **graph** may combine data from several pairs or from the whole class.

Survey Graphs

If your students are unfamiliar with graphing, you can give them some practice by having them make **survey graphs**. Students survey all members of the class (or the school or some other group) on a particular question and then represent the data they have gathered. For example, students might conduct surveys to find out the favorite ice cream flavor, favorite music group, how students get to school, or favorite novel. You and your students will surely come up with other ideas to use for survey graphs.

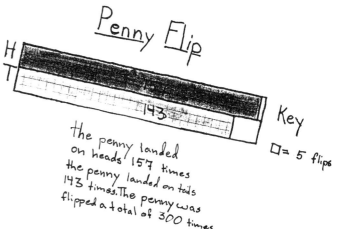

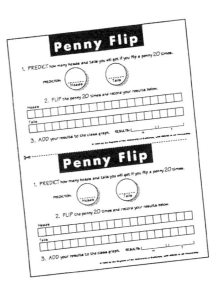

Wide rolls of graph paper are available from some teaching supply stores and catalogs. Easel pads with graph paper are available at some office supply stores.

Suggest that students limit the choices they give when they conduct their survey. For example, they may give three or four flavors of ice cream and ask each student to choose their favorite among those. Otherwise, if they leave the choice of favorite ice cream open-ended, they may end up with thirty kids choosing thirty flavors. That would make a large graph that does not lend itself to much interpretation.

Data Sheets

Data sheets, used to record data gathered by a pair of students, are included in this guide. Many teachers choose to have students make their own data sheets instead. A data sheet can be as simple as using tally marks on a piece of scratch paper. Alternatively, a data sheet can be carefully planned and neatly drawn on graph paper.

Graphing

Black line masters of graph paper are included in the guide for students to use when designing their own graphs. Students benefit greatly by organizing data themselves and communicating the information to others in a succinct and "graphic" fashion. Logic and reasoning skills are called upon, and various problems and choices often need to be addressed. For example: How could we make a graph if the class got 193 heads when we flipped the penny, and the graph is only 25 squares high?

Allow plenty of time for students to make their own graphs. Students may need a full class period or more just to make their graphs. This time is far from wasted, as students gain valuable skills in constructing and interpreting graphs.

Suggestions for making graphs are found in the text of each activity. These, however, should serve only as general guidelines, as you are in the best position to decide how to effectively instruct and challenge your students. Below are some options for graphing that other teachers have found successful.

- Younger children can combine their results with those of one or two other pairs and graph that data, rather than graphing the data from the entire class.

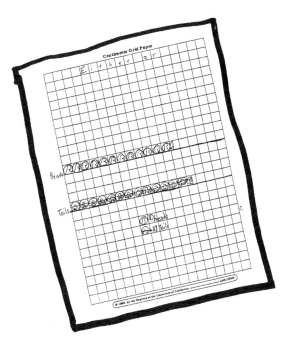

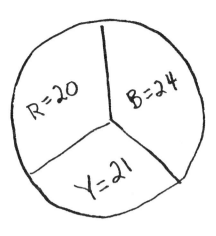

- Have students represent the data on blank paper rather than graph paper. Students can use their creativity to devise ways other than traditional graphs to represent the information.

- Have students make a pie chart.

- Have students use a tree diagram.

- Have students use symbols to represent data. This can help them see patterns.

- Lead a class discussion about different ways of graphing and representing data, using graphs created by the students or clipped from newspapers and magazines. A rich discussion of various graphing approaches can result. For example, students may discover that certain methods of recording data make comparison between results simpler, or easier to read, etc.

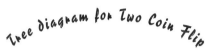

Tree diagram for Two Coin Flip

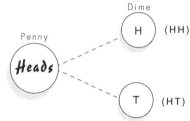

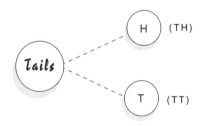

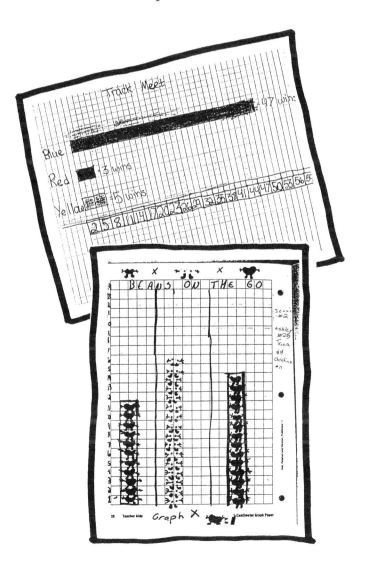

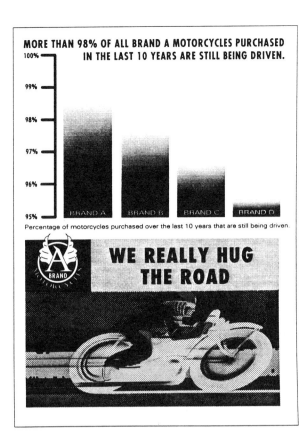

MORE THAN 98% OF ALL BRAND A MOTORCYCLES PURCHASED IN THE LAST 10 YEARS ARE STILL BEING DRIVEN.

Percentage of motorcycles purchased over the last 10 years that are still being driven.

WE REALLY HUG THE ROAD

• The teacher can model one method of making a graph on the overhead.

• Have students collect and analyze graphs from newspapers and magazines. These can provide a rich source for ideas in graphing. Students can collect samples of graphs. In Activity 2: Track Meet, students look at variations in presentation style, as well as the use of graphics, large type, and colors. Discussing and displaying a variety of ways to represent data can inspire creativity in students when they design their own graphs. Older students might notice how the visual presentation of a graph affects the meaning or possible interpretation of the information.

Who Uses Statistics and Probability?

Children meet probability in many of the games they play, especially those using dice or spinners. They see statistics on baseball cards, cereal boxes and TV news shows and weather reports.

As adults, probability is involved in games we play, weather predictions and economic forecasts. We use statistics to help decide what new car to purchase, which food are lower in fat than others, and whether our candidate is likely to win.

Many other careers involve various aspects of statistics and probability. Some careers that emphasize probability and statistics are in accounting, the stock market, insurance, real estate, public opinion polling, urban planning, advertising, psychology, elementary school teaching, public health, meteorology (the study and prediction of weather and climate), epidemiology (the study of epidemics, which involves both gathering data and predicting the future course of outbreaks of illness), and seismology (the study of earthquakes, which includes using historical as well as geological data to predict future quakes). There are elements of statistics and probability in many other professions.

Writing and Mathematics

Some teachers have been using writing in conjunction with mathematics for many years. Others have begun more recently to integrate writing as a great way to help students develop mathematically. Writing in mathematics can help students formulate their ideas, give the teacher information about student understanding, and provide an excellent means of assessing student progress.

At the end of each activity in this guide are several writing suggestions. Some are very specific in nature, such as to describe the results you and your partner got. Other suggestions are more open-ended, for example—which spinner did you like using the most and why? You (and your students!) may come up with other ideas for writing as well.

10-14-92

You flip the penny 20 times and you have a 50/50 chance of getting heads or tails. Heads came up more than tails in some cases. The first day we did this, the results were 48 tails and 52 heads. The next day We made a chart and averged the numbers of heads and tails of the hole class (Heads 139, Tails 121). We also had an example of probability. He said When a kid woke up one morning for school and the power was off. He went over to his sock drawer. There were black and white all mixed up in the drawer. The question was how many Socks does the kid have to pick to be gaurentied a pair of socks (My teacher also said it's pitch black in the room. He'll take the socks outside to see his pair.) The answer was 3.

(Track Meet)

Spinner X
Blue is the winner!

The blue won in ours. My partner was Kristina. I voted for blue too. We played two games. Blue won both times.

Spinner Z
Blue is the winner)

Blue won three times. My partner is still Kristina. This time I thought blue would win too. Blue had more chance this time because half the circle was blue and only a quarter of red and yellow were on the spinner. I predicted blue would win, and it did!! It did not suprize me that blue won!!

Although writing does require an additional time commitment, teachers have found it very helpful and valuable to integrate the use of writing in mathematics. Some daily or weekly writing assignments can relate to mathematics, rather than another subject area. This can help students see that writing is used for a purpose, whether that purpose is to consolidate thinking, save data or information for later use, or communicate with others. The strong connections with the language arts are evident, and there are many creative ways to combine writing assignments, mathematics, and literature connections (such as those listed in this guide) and thereby reach across many parts of the curriculum. Exploring related concepts through playing a game, writing about it, creating a story, or reading a book has a special "multiplying effect" and can more firmly establish a concept in our students' (and our own) minds.

The Probability in the Activities

This background information is included for teachers only, and is not meant to be read out loud to students, or handed out, or used as a checklist of what they "should" learn. **The information in this section is generally geared at an *upper* elementary level.** Depending on the age and experience of your group, you can decide what information from this section you wish to introduce, and how best to do so.

The Law of Large Numbers

You may notice that when your students flip their pennies 20 times each, there is quite a range of results. Probably, few or none of your students will get ten heads and ten tails. Since the probability of getting heads is 1 out of 2, it would seem logical to get 10 out of 20. However, a small sample (such as 20 flips) may give results that differ quite a bit from the theoretical probability.

If the sample is large enough, the results are likely to be a close approximation of the theoretical probability. For example, if you flip a penny 1000 times, you will almost certainly get close to 500 heads and 500 tails. Older students may start to see this phenomenon on their own as they participate in the activities of this unit: their own results (a small number of trials) may not be close to the probability while the class results (a larger number of trials) are likely to be quite close. Some students may notice that the class data more closely reflects the probability than the data from individual pairs of students.

The "law of large numbers" says that a large number of trials will give results very close to the theoretical probability. A small number of trials may or may not give results very close to the theoretical probability.

Is "Almost" Good Enough?

If your class gets 193 heads and 227 tails during the penny flip activity, they may think their pennies are not behaving as predicted, that they are "loaded" or lucky. Developmentally, some children may not be ready to understand that **almost** half heads and half tails really is following the laws of probability. Students will understand the activities at their own developmental level.

In talking to your students, you can use phrases such as "almost the same number of heads as tails," "very close to half heads and half tails," or "nearly 50% heads" to help students begin to see beyond the fact that the numbers are not **exactly** equal.

Activity 1: Penny Flip

In Activity 1, students flip a penny and record the results. In the case of flipping one penny, there are two possible outcomes: the penny will land on heads or it will land on tails. It is so unlikely that the penny would stay in the air or land on its edge, that we don't need to consider either of these a possibility. The penny is "fair"—when you flip a penny, it is **equally likely** that it will land on heads as on tails.

In other words, we can say that there are two equally likely outcomes every time you flip a coin. If you are hoping that the penny lands on tails, you have a **one out of two chance** of getting your wish. The "one" is the result you want (tails, in this case) and the "two" is the total number of equally likely outcomes (heads and tails).

This can be written **symbolically** as a fraction or ratio (1/2). You can represent probability as a decimal (0.5) or a percent (50%). We can also say there is a 50/50 chance of getting tails. The same can be said for getting heads.

What is the probability of getting either heads or tails on your next flip? 100%. A probability of 100% means that something will definitely occur, and you will certainly get either heads or tails next time you flip that penny.

What is the chance of getting a quarter when you flip a penny? 0%. A probability of 0% means that an event will never happen. Your penny will not turn into a quarter no matter how many times you flip it.

Activity 2: Track Meet

In Activity 2, students first play the game using Spinner 1, which is divided into three sections of equal area. This is another example of a situation with **equally likely probabilities**. There is a one in three probability of getting yellow. The same is true of both red and blue. This can be represented numerically. There is a one in three chance of getting yellow (1/3). The chance of getting yellow is about 33% or 0.33.

Students then play a rematch with either Spinner 2 or Spinner 3. Although these spinners look different, they both carry the same probability because, yellow and red each cover one quarter of the spinners and blue covers one half . This is the first activity in which the students experience **unequal chances**. There are three possible outcomes, but they are not equally likely.

With both Spinners 2 and 3, red will come up about 25% of the time, yellow will come up about 25% of the time and blue will come up about 50% of the time.

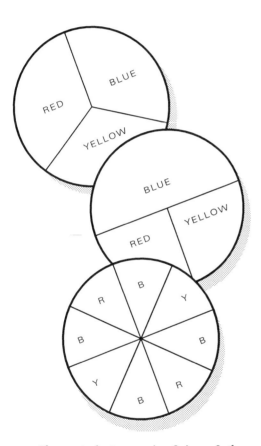

If your students are using Spinner 3, they will notice that it is divided into eight equal sections. It is equally likely that the spinner will land in any one of those eight sections. Since two sections are yellow, there is two out of eight probability that any one spin will land on yellow. Since 2/8 = 1/4, there is a 25% chance that any one spin will land on yellow. The same is true of red. Since four of the sections are blue, there is a four out of eight probability that any one spin will land on blue. Since 4/8 = 1/2, there is a 50% chance that any one spin will land on blue.

For Spinners 2 and 3, the probability can be written **symbolically** as a fraction or ratio (1/4), as a decimal (0.25), or as a percent (25%). The same is true for red. For blue, the probability can be expressed as 1/2, 0.5, or 50%.

Activity 3: Roll a Die

In Session 3, students investigate the probability of getting the numbers one through six when rolling a die. One die presents six possible choices, and, as long as the die is not weighted unfairly, each of the numbers one through six has an **equally likely chance** of landing up on any one toss. With a large number of tosses, each number will most likely be rolled about the same amount of times as every other number.

This activity presents another good opportunity to help students see the law of large numbers in action. Results from pairs of students are likely to vary greatly from the theoretical probability (that each number will occur about one sixth of the time.) Class results will be closer to the theoretical probability. If another class is doing the same activity, combining the results of both classes can further illustrate the law of large numbers.

Session 4: Horse Race

In Session 4, students use two dice to play the game, and the winning horses are recorded on a large class graph. In contrast to the equally likely chances of getting the numbers one through six with one die, with two dice the chances of getting each possible sum (two through twelve) are **not equally likely**. The class graph of the winning horses creates a graph, and students can see that it was not a "fair" race: some horses won many times and other horses won only a few times or not at all. The class graph of winning horses frequently looks like a **bell shaped curve**.

The "Keeping Track" chart can help students understand why certain horses won so much more often. "Keeping Track"represents the 36 possible ways for two dice to land when tossed together. The sum for each of the 36 tosses is displayed on the chart.

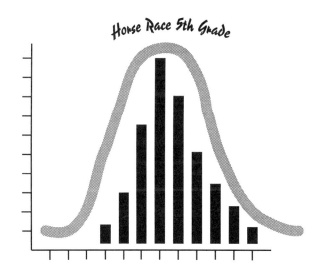

*When playing the "Race Track" game, the **combinations** or sums of the numbers on the dice are important. When using the "Keeping Track" chart to look at probabilities, it is the **permutations**, or different ways the sums are made, that are important.*

From the chart, you can see that there is only one way to get a sum of two using two dice (1+1), so a two has a one out of 36 chance of coming up on any one toss (1/36). There are six ways to get a sum of seven, so a seven has a six out of 36 probability of coming up on any one toss (6/36 or 1/6). The probability for the other sums can be figured out in the same way.

The sums six and eight can each occur five ways, so probability predicts that they will occur nearly as many times as the sum seven does. Occasionally, either Horse #6 or #8 will win more times than Horse #7. That does not mean something is wrong with the dice or that the students are cheating! Probability says what will probably happen, not what will definitely happen.

Session 5: Native American Game Sticks

In Session 5, students design their own sticks and play a version of a gambling game played by Native Americans in California and in many other parts of North Ameerica.

As in Horse Race, described above, when the game is played, it is the **combinations** (of the sticks in this case) that are important—counters are taken according to the combination of design sides and plain sides. When examining the probability for Game Sticks, one must look at the **permutations**, or all the possible ways the sticks can land.

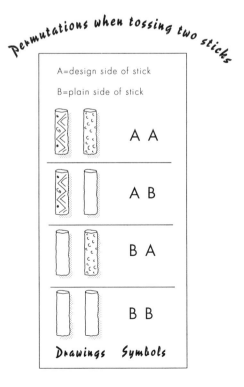

Permutations when tossing two sticks

A=design side of stick

B=plain side of stick

A A

A B

B A

B B

Drawings   Symbols

To gain some insight into the probability of the Game Sticks, it helps to first make things simpler. Begin with just one stick. When one stick is tossed, there are two possible outcomes, as in the Penny Flip. The stick may either land with the design side up or the plain side up. (With the penny, it was either heads or tails.)

When two sticks are tossed, there are three combinations: two design sides, two plain sides, or one of each. However, there are four permutations, as the illustration shows.

For two sticks, the probability (based on the permutations) of getting two design sides is one out of four. The probability is the same for getting two plain sides, one out of four. However, for getting one of each, the probability is two out of four.

For four sticks, the sticks can fall in 5 combinations. However, there are sixteen permutations. Notice that there is a one out of sixteen chance of getting all design sides up, but a 6 out of sixteen chance of getting two designs and two plain.

The Game Sticks activity, which uses six sticks, presents an even more unwieldy problem. As you can see by the chart on the next page, it is a difficult problem to diagram. Although there are only seven combinations, there are 64 permutations! Examining the probabilities involved in Game Sticks is a problem that might be posed to students at the high school or college level, and would be solved with the aid of probability formulas.

The second session of Game Sticks suggests a method of helping students to begin seeing the difference between combinations and permutations. Depending on your students skills and interest, you many decide to do all, some or none of that section. Remember, it is not expected that third through sixth grade students will master all the complex probability of this game, but rather that they will gain concrete experiences with probability in a variety of circumstances.

Combinations and Permutations for Native American Game Sticks

■	A = Design Side
□	B = Non-Design Side

TWO STICKS

Combinations	Permutations
■ ■	AA
■ □	AB
□ ■	BA
□ □	BB

FOUR STICKS

Combinations	Permutations
■ ■ ■ ■	AAAA
■ ■ ■ □	AAAB
	AABA
	ABAA
	BAAA
■ ■ □ □	AABB
	ABAB
	ABBA
	BAAB
	BABA
	BBAA
■ □ □ □	ABBB
	BABB
	BBAB
	BBBA
□ □ □ □	BBBB

SIX STICKS

Combinations	Permutations
■ ■ ■ ■ ■ ■	AAAAAA
■ ■ ■ ■ ■ □	AAAAAB
	AAAABA
	AAABAA
	AABAAA
	ABAAAA
	BAAAAA
■ ■ ■ ■ □ □	AAAABB
	AAABAB
	AABAAB
	ABAAAB
	BAAAAB
	AAABBA
	AABABA
	ABAABA
	BAAABA
	AABBAA
	ABABAA
	BAABAA
	ABBAAA

SIX STICKS (continued)

Combinations	Permutations
	BABAAA
	BBAAAA
■ ■ □ □ □ □	AABBBB
	ABABBB
	ABBAB B
	ABBBAB
	ABBBBA
	BAABBB
	BABABB
	BABBAB
	BABBBA
	BBAABB
	BBABAB
	BBABBA
	BBBAAB
	BBBABA
	BBBBAA
■ □ □ □ □ □	ABBBBB
	BABBBB
	BBABBB
	BBBABB
	BBBBAB
	BBBBBA
□ □ □ □ □ □	BBBBBB
■ ■ ■ □ □ □	AAABBB
	AABABB
	AABBAB
	AABBBA
	ABAABB
	ABABAB
	ABABBA
	ABBAAB
	ABBABA
	ABBBAA
	BAAABB
	BAABAB
	BAABBA
	BABAAB
	BABABA
	BABBAA
	BBAAAB
	BBAABA
	BBABAA
	BBBAAA

Native American Game Sticks

Many Native American cultures include games of chance in their diverse recreational activities. One common variety throughout the Americas involves dice, in which the numbers are determined by throwing objects with two faces. The game pieces can be made from a variety of materials, including split canes, wood, bones, beaver or woodchuck teeth, walnut shells, peach and plum stones, and sea shells.

There are two main methods used to keep track of the score: one in which the score is kept with counters that pass from hand to hand, and the other in which a counting board or abacus is used. Both men and women play dice games, though usually separately.

One well-known game is Gambling Sticks, after which the Game Sticks game in Activity 5 is modeled. Many of the designs featured on the "Traditional Designs from California" sheet are drawings of designs originated by the Pomo Indians of California. This game was traditionally played by Native American women who lived in what is now the western portions of the United States, Mexico and Canada. The women played especially during the wet winter months when they were not as busy with other responsibilities. Items of value would be gambled. For example those living on the coast might gamble shells, while those living in the mountains might wager obsidian.

Each woman would have her own set of six sticks to play the game. The sticks were marked with designs on one side only. In turn, each player tossed her sticks, either from a basket or by hand. Depending on the combination of sides in which the sticks landed, players obtained counters. The object of the game was to obtain all ten counters that were used to keep score. Each woman brought an item to bet. The woman who obtained all ten counters won the item that the other person wagered. As the games were played, songs and chants could often be heard to help bring luck to the players.

Resources

Games of the Native American Indians by Stewart Culin. Dover Publications, New York, 1975.

Exporing Probability by Claire M. Newman, Thomas E. Obremski, and Richard L. Scheaffer. Dale Seymour Publications, Palo Alto, California, 1987.

Games of the Native American Indians by Stewart Culin. Dover Publications, New York, 1975.

Hands-On Statistics, Probability, and Graphing (K-9) by Linda Sue Brisby et al. Hands On Inc., Solvang, CA 1988 (available from Activity Resources, Hayward, California).

How to Lie With Statistics by Darrell Huff. Dale Seymour Publications, Palo Alto, California.

NCTM 1981 Yearbook, Teaching Statistics and Probability by National Council of Teachers of Mathematics, Reston, Virginia.

Probability Jobcards: Intermediate (Gr. 3-6) by Shirley Hoogeboom and Judy Goodnow. Creative Publications, Mountain View, California, 1977.

Probability Model Masters by Dale Seymour. Dale Seymour Publications, Palo Alto, California.

Quadice by Elizabeth Stage et al., Great Explorations in Math and Science (GEMS) teacher's guide for Grades 4–8. An original game involving four dice, which introduces basic probability concepts. Includes a cooperative version of the game.

Socrates and the Three Little Pigs by Tuyosi Mori. Philomel, New York, 1986.

Statistics by Jane Srivastava. Harper, 1973.

Used Number Series:
(available from Dale Seymour Publications, Palo Alto, California)
> *Statistics: The Shape of the Data* by Susan Jo Russell and Rebecca B. Corwin, 1989.
> *Statistics: Prediction and Sampling* by Rebecca B. Corwin and Susan N. Friel, 1990.
> *Statistics: Middles, Means, and In-Betweens* by Susan N. Friel, Janice R. Mokros, and Susan Jo Russell, 1992.

What Are My Chances, Book A (Grades 4-6) by Albert P. Shulte and Stuart A. Choate. Creative Publications, Mountain View, California, 1991.

Winning With Numbers: A Kid's Guide to Statistics by Manfred G. Riedel. Prentice Hall, 1978.

Literature Connections

Books in which statistics are generated often naturally involve collecting, organizing, and recording data. Though some of these books may not be at the exact reading level of your students, they can still provide the basis for a good discussion or a statistics activity. In addition, books about prediction, chance and probability as well as number are included. The book of Native American legends is an accompaniment to the "Game Sticks" activity in Session 5. For a much more comprehensive listing of books on Native American themes see the GEMS guide *Investigating Artifacts* and/or *Once Upon A GEMS Guide*, the GEMS literature connections handbook, especially under the Math Strands of Probability & Statistics, Logic, Number, and Pattern. Stories that center around playing a game that involves probability and that include some controversy about the chances of winning make excellent connections. Non-fiction books that focus on careers or issues that involve the study and analysis of statistics would also make excellent extensions. We welcome your suggestions!

Alice
by Whoopi Goldberg; illustrated by John Rocco
Bantam Books, New York. 1992
Grades: 2–6
This book not only highlights the comedic skills of its author, it contains a lesson about friendship and some statistical wisdom relating to sweepstakes and their deceptive enticements. Alice enters "every sweepstakes, every giveaway, every contest," because she wants more than anything else to be rich. She lives in New Jersey and one day is notified she has won a sweepstakes. Alice convinces her friends to go with her on an odyssey to New York City to collect the prize. What happens makes for a rollicking adventure, which your students will enjoy, at the same time as they realize the probability and statistics lessons they are learning in the classroom have lots of applications in the real world!

Back in the Beforetime: Tales of the California Indians
retold by Jane L. Curry; illustrated by James Watts
Macmillan Publishing Co., New York. 1987
Grades: 2–6
A retelling of 22 legends about the creation of the world from a variety of California Indian tribes. In the myth "The Theft of Fire," the animal people spend an evening gambling with the people from the World's End. After the animal people lose all they can gamble, Coyote wagers the animal people's fire stones in a final bet. The outcome of that bet is the basis for the mythic explanation of how the animal people got fire. The story ties in with Session 5.

Cloudy With a Chance of Meatballs

by Judi Barrett; illustrated by Ron Barrett
Atheneum, New York. 1978
Grades: K–3

A hilarious look at weather conditions in the town of Chewandswallow, which needs no food stores because daily climactic conditions bring the inhabitants food and beverages, such as a storm of giant pancakes or an outpouring of maple syrup. This book presents a non-threatening way to look at predictions. Students can follow up the story by listening to weather reports and charting the accuracy of meteorologists. They can also use *The Cloud Book* by Tomie dePaola to observe and chart clouds, one aspect of weather patterns.

Jumanji

by Chris Van Allsburg
Houghton Mifflin, Boston. 1981
Scholastic Books, New York. 1988
Grades: K–5

A bored brother and sister left on their own find a discarded board game (called Jumanji) which turns their home into an exotic jungle. A final roll of the dice for two sixes helps them escape from an erupting volcano. The story complements the horse racing game in this guide where the roll of the dice also determines an important outcome.

People

by Peter Spier
Doubleday, New York. 1980
Grades: Preschool–6

Here's an exploration of the differences between (and similarities among) the billions of people on earth. It illustrates different noses, different clothes, different customs, different religions, different pets, and so on. This is a great book to use in collecting statistics and creating graphs about characteristics of people. Pairs of students can investigate the occurrence in their class of a physical feature (hair type, eye color, etc.), preference (types of food), or other distinguishing attribute (where one lives), and report their findings to the class.

Summary Outlines

Activity 1: Penny Flip

Getting Ready
1. Discuss making predictions with students as needed.
2. Decide how you want students to record data.
3. Have students begin collecting graphs from newspapers and magazines.

Session 1: Heads or Tails?
Introducing the Penny Flip
1. Have students share knowledge about flipping coins.
2. Ask students to predict how many heads/tails if coins flipped 20 times, and explain why they think so.
3. Have a student flip penny. Demonstrate recording results on data sheet.
4. Stop several times for class to discuss results.
5. Ask students for "true statements" about the graph.

The Penny Flip
1. Have students work in pairs to flip penny 20 times and record on data sheet.
2. Invite students to report data to the class.
3. You may want to have students flip 20 more times for more data.

Session 2: Graphing and Class Conclusions
Graphing and Discussing
1. Review Penny Flip data sheet.
2. Survey students and take predictions on entire class results.
3. Gather and record data from all groups.
4. Elicit graphing suggestions.
5. Have students create graphs.
6. Discuss possible reasons for the results.
7. Predict—how many heads and tails if 100 flips?
8. With older students, discuss concept/theory of *probability*, the phrase **"equally likely chance."** Probability tells us approximately, but not exactly, what will happen. You could introduce symbolic ways for writing probabilities.
9. Consider writing extensions as described in guide.

Activity 2: Track Meet

Getting Ready
1. Decide if students will use Spinner 2 or 3.
2. Have students make spinners according to instructions in guide.
3. Duplicate board game, make transparencies, prepare two class graphs.

Session 1: Track Meet: Spinner 1

1. Ask students other games they know that have spinners.
2. Display Spinner 1 and explain "Track Meet" game.
3. Survey students on expected outcome.
4. Play demonstration game with a student, pausing for comments.
5. Record the winner on the class data graph.
6. Discuss need for accurate recording of results.

Collecting and Reporting Data

1. Have students make predictions at start of every game.
2. Students play several games and record winners on class graph.
3. Afterward, ask students for comments, surprises.
4. Ask for "true statements" about class graph.
5. Display spinner on overhead; ask about relative divisions of colors.
6. Encourage older students to notice connections between spinner and results.

Session 2: Track Meet Rematch (Spinner #2 or #3)

Preparing for the Rematch

1. Show new spinner with #1 on overhead.
2. How are spinners same? How different?
3. Remind students to make predictions.

Playing the Game

1. Allow time for several games.
2. Discuss results; compare to penny flip and previous "Track Meet" game. Discuss "equally likely chance."

Session 3: Graphing and Class Conclusions

Exploring Graphs from Newspapers and Magazines

1. Each pair of students gets two or more newspaper/magazine graphs and chooses one to find out more about.
2. After few minutes, pairs join to make groups of four.
3. Each pair explains graph to other pair.
4. Groups report to entire class on what they found out about how data can be represented.

Graphing Track Meet Data

1. Post class graphs for both spinners and review results.
2. Pairs make graphs: each doing one of the two "Track Meets."
3. Circulate, encouraging discussion and comparison.
4. Have each pair record similarities and differences between their two graphs.
5. Post the graphs. Compare similarities and differences.
6. How might students explain differences in results between the two spinners? How do the results relate to the spinners?

Comparing the Spinners

1. Display both spinners on overhead.
2. How much of the circle do the colors cover?
3. Discuss "fairness." Was one spinner more "fair" than the other?
4. Have students imagine themselves as "blue runner" or "yellow runner."
5. For second spinner, does one color have a greater chance of coming up?
6. Does either spinner represent equally likely chance?
7. Consider writing extensions.

Activity 3: Roll a Die

Getting Ready

Duplicate one data sheet or graph paper for each pair of students.

Session 1: Gathering Data

Rolling a Die

1. Share knowledge about dice.
2. When rolling standard die, will one number come up more?
3. Predict how many of each number come up in 30 rolls.
4. Display data sheet on overhead.
5. Volunteer rolls die 30 times while you record results.
6. Ask for "true statements" and predictions.

Collecting and Reporting the Data

1. Students roll die and record data.
2. A few students report to class. Is a die fair or unfair?
3. Save data sheets for next session.

Session 2: Graphing and Class Conclusions

1. Hold brief class discussion on results.
2. Post results from each pair for all.
3. Have students calculate class totals and report.
4. Students make graphs.
5. Post graphs—ask for "true statements."
6. What connection between dice and data on graph?
7. Show older students numerical ways to represent probability for one die.
8. Consider writing extensions.

Activity 4: Horse Race

Getting Ready

1. Duplicate one copy of game board and one "Keeping Track" sheet for each pair of students.
2. Make a class graph.

Session 1: Racing the Horses

A Day at the Races

1. Students review results of last activity.
2. Today's game uses two dice.
3. Display game board on overhead and explain that the horse whose number is the sum of the dice moves ahead **ONE** space.
4. Have students predict which horse will win.
5. Demonstrate game on overhead with two student volunteers.
6. Pause frequently to consider results.
7. Post the class graph and record winning horse.

Playing the Game

1. Remind students about accurate recording.
2. Give out game boards, counters, dice.
3. Each race recorded after it's over.
4. Discuss results: "true statements," any surprises?

Session 2: How Many Combinations?

All Possible Ways

1. Post the class graph from previous session, review results.
2. How many ways can you make a 7 with two dice?
3. **List ways you** can make several other numbers.

"Keeping Track" Chart

1. Display chart on overhead.
2. Color dice on transparency
3. Give chart to students and have them color the dice to match the ones they used.
4. Students fill out chart and see how many ways to make sums from 2 to 12.
5. Ask how the chart relates to "Horse Race" game.
6. Older students can figure out fractional/decimal probabilities.
7. Consider writing extensions.

Activity 5: Native American Game Sticks

Getting Ready

1. Read over information on game.
2. Make copy of "Traditional Designs" for each pair of students.
3. Use 12 tongue depressors and colored pens to make two sets of game sticks.
4. Consider posting/displaying rules of game.

Session 1: Making and Playing Game Sticks

Demonstrating the Game
1. Game similar to games played by Native Americans.
2. Post and explain the rules as shown in guide.
3. Divide class into two groups, put out 10 counters.
4. Team with most design sides goes first.
5. Play game until one team has all 10 counters. with students alternating so all get chance to toss game sticks.
6. Stress difference between "toss" and "throw."
7. Teams could sing/chant encouragement (no jeers allowed).

Making Game Sticks
1. Display traditional designs on overhead.
2. Remind students to decorate only one side and write name or initials on other side.
3. Give out sticks, traditional designs, colored markers to each pair of students.
4. Students decorate their game sticks.

Play the Game
1. Students who finish early can play with others who are finished.
2. Pairs or small groups play the game.

Play It Again
1. In another class session, play more, in pairs or small groups.
2. Ask students for comments.
3. Refer to rules—why were they made the way they were?

Session 2: A Probability Experiment
How Likely?
1. Discuss how the sticks tended to fall.
2. Describe ways sticks could land.
3. Ask: Are some ways more likely than others?
4. What does probability have to do with Game Sticks?

The Experiment
1. Consider simpler versions of game to learn more.
2. How many ways can one game stick land? How is that like the penny flip? How is the probability similar?
3. Consider tossing two game sticks.
4. Students work in pairs to record possibilities for one, two, and four sticks.
5. What did students discover about how four sticks can land?
6. What is their prediction for all six sticks? How is probability incorporated into the rules?
7. Have class collect data for six sticks and put on class graph.
8. Students describe graph and explain why they think they got this data.
9. Consider writing extensions as described in guide.

GEMS Guides

Please contact GEMS for a descriptive brochure and ordering information

TEACHER'S GUIDES

Acid Rain
Grades 6–10

Animal Defenses
Preschool–K

Animals in Action
Grades 5–9

Bubble Festival
Grades K–6

Bubble-ology
Grades 5–9

Buzzing a Hive
Preschool–3

Chemical Reactions
Grades 6–10

Color Analyzers
Grades 5–8

Convection: A Current Event
Grades 6–9

Crime Lab Chemistry
Grades 4–8

Discovering Density
Grades 6–10

Earth, Moon & Stars
Grades 5–9

Earthworms
Grades 6–10

Experimenting with Model Rockets
Grades 6–10

Fingerprinting
Grades 4–8

Frog Math: Predict, Ponder, Play
Grades K–3

Global Warming
Grades 7–10

Group Solutions
Grades K–4

Height-O-Meters
Grades 6–10

Hide a Butterfly
Preschool–3

Hot Water & Warm Homes from Sunlight
Grades 4–8

In All Probability
Grades 3–6

Investigating Artifacts
Grades K–6

Involving Dissolving
Grades 1–3

Ladybugs
Grades Preschool–1

Liquid Explorations
Grades K–3

Mapping Animal Movements
Grades 5–9

Mapping Fish Habitats
Grades 6–10

Moons of Jupiter
Grades 4–9

More Than Magnifiers
Grades 6–9

Of Cabbages & Chemistry
Grades 4–8

Oobleck: What Do Scientists Do?
Grades 4–8

Paper Towel Testing
Grades 5–8

QUADICE
Grades 4–8

Terrarium Habitats (late 1993)
Grades K–6

Tree Homes
Grades Preschool–1

River Cutters
Grades 6–9

Vitamin C Testing
Grades 4–8

ASSEMBLY PRESENTER'S GUIDES

The "Magic" of Electricity
Grades 3–6

Solids, Liquids, and Gases
Grades 3–6

EXHIBIT GUIDES

Shapes, Loops & Images
all ages

The Wizard's Lab
all ages

HANDBOOKS

GEMS Teacher's Handbook

GEMS Leader's Handbook

A Parent's Guide to GEMS

Once Upon a GEMS Guide
(Literature Connections to GEMS)

To Build a House
(Thematic Approach to Teaching Science)

Write or Call:
GEMS
Lawrence Hall of Science
University of California
Berkeley, CA 94720
(510) 642-7771

Penny Flip

1. PREDICT how many heads and tails you will get if you flip a penny 20 times.

PREDICTION:

_____ Heads   _____ Tails

2. FLIP the penny 20 times and record your results below.

Heads

Tails

3. ADD your results to the class graph.   RESULTS: (_____)_(_____)
                                                        H   T

✂ -

Penny Flip

1. PREDICT how many heads and tails you will get if you flip a penny 20 times.

PREDICTION:

_____ Heads   _____ Tails

2. FLIP the penny 20 times and record your results below.

Heads

Tails

3. ADD your results to the class graph.   RESULTS: (_____)_(_____)
                                                        H   T

Spinner 1

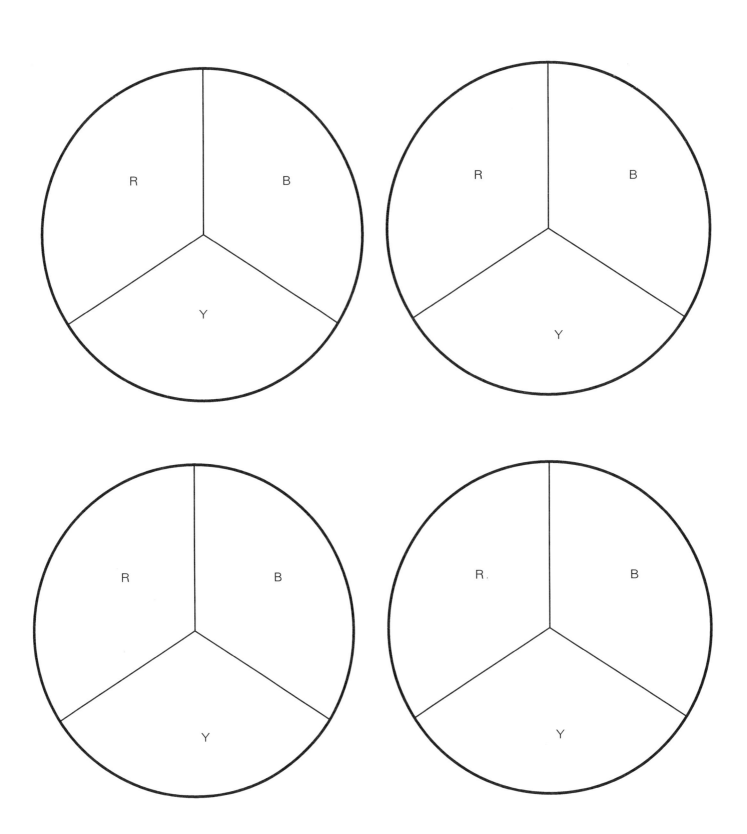

Spinner 2

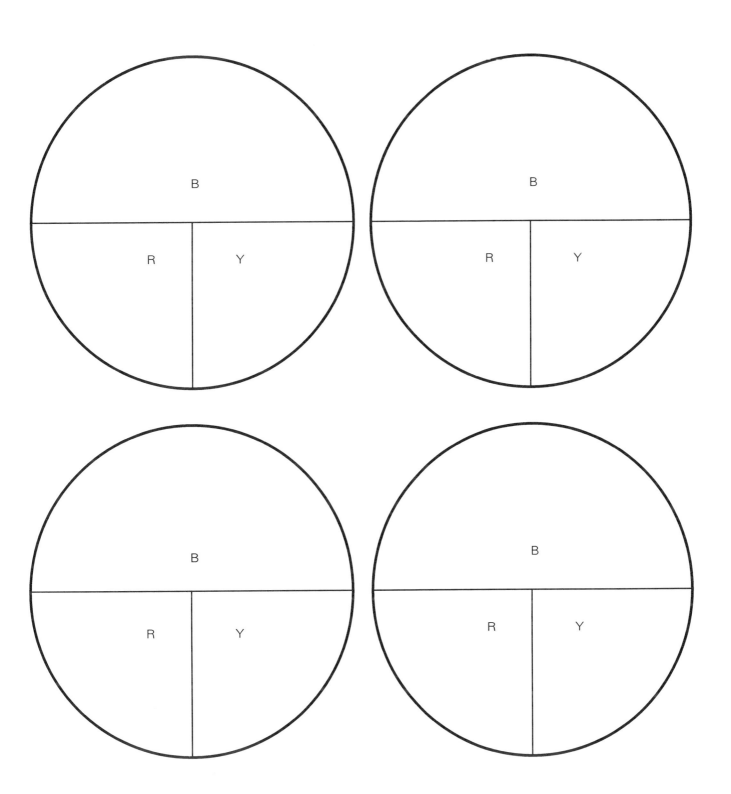

Spinner 3

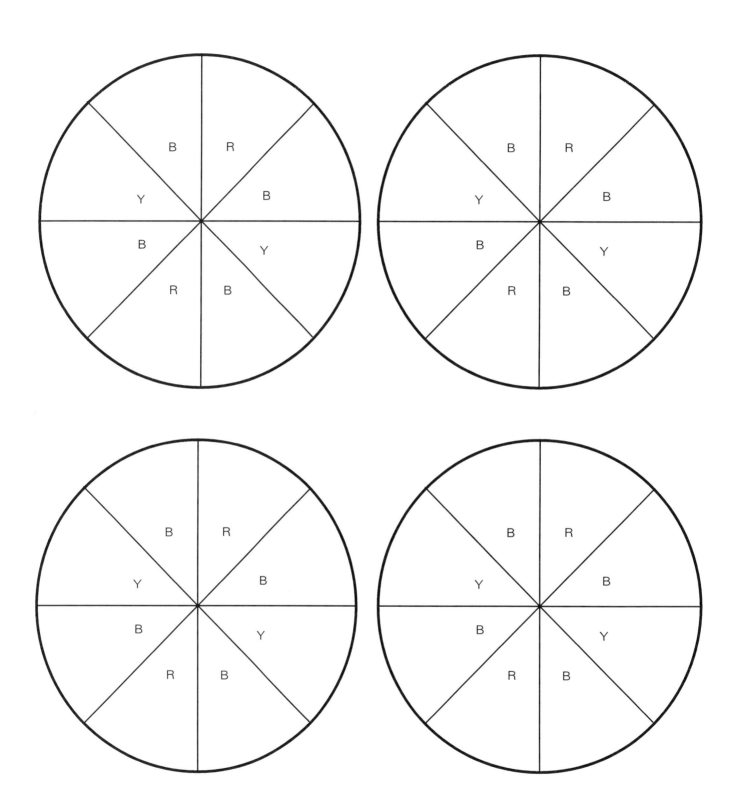

T R A C K   M E E T

Finishing line

Starting line

YELLOW   **BLUE**   **RED**

Roll a Die

DATA SHEET

- Roll the die **30** times.

- Record your results below.

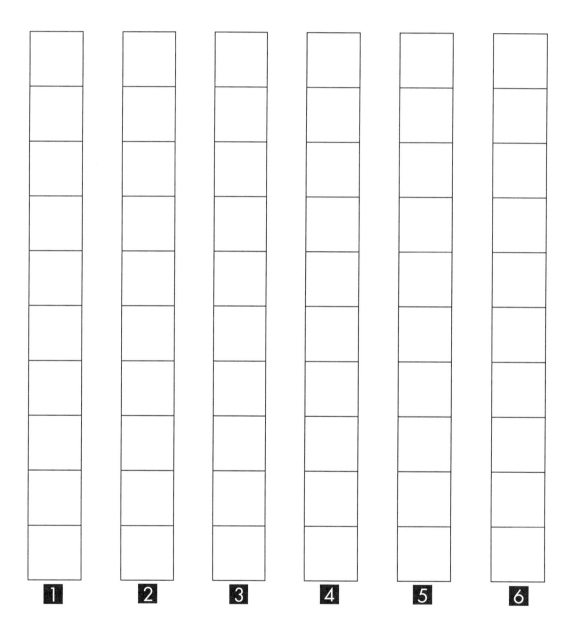

| 1 | 2 | 3 | 4 | 5 | 6 |

TOTAL ——— ——— ——— ——— ——— ———

- Add your results to the class graph.

Keeping Track

TRADITIONAL DESIGNS
from CALIFORNIA

Native American Game Sticks

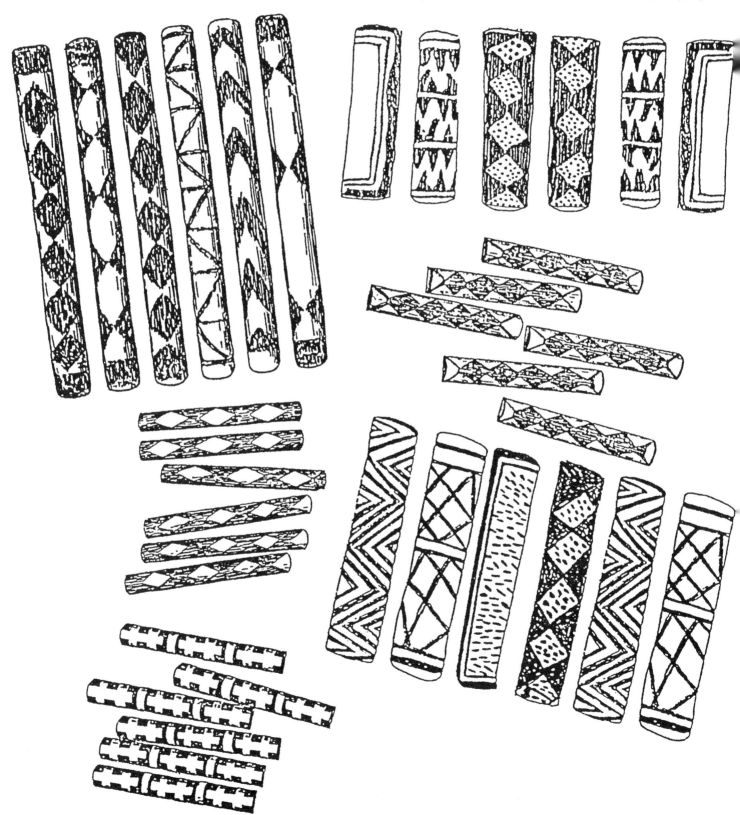